Time Lord

Clark Blaise

TIME LORD

THE REMARKABLE CANADIAN WHO MISSED HIS TRAIN AND CHANGED THE WORLD

Alfred A. Knopf Canada

To John and Myrna Metcalf

PUBLISHED BY ALFRED A. KNOPF CANADA

Copyright © 2000 by Clark Blaise

All rights reserved under International and Pan-American Copyright
Conventions. Published in 2001 by Alfred A. Knopf Canada, a division of
Random House of Canada Limited, Toronto, and in the United States by Pantheon
Books, a division of Random House, Inc., New York. First published in Great Britain
by Weidenfeld & Nicolson, an imprint of Orion Publishing Group Ltd., London,
in 2000. Distributed by Random House of Canada Limited, Toronto.

Knopf Canada and colophon are trademarks.

Canadian Cataloguing in Publication Data

Blaise, Clark, 1940–
Time lord : The remarkable Canadian who missed his train and changed the world
ISBN 0-676-97252-7
1. Fleming, Sandford, Sir, 1827-1915.
2. Time—Systems and standards—History. I. Title.

QB223.B52 2001   389.17′09   C00-932414-3

First Canadian Edition

Visit Random House of Canada Limited's Web site: www.randomhouse.ca

Printed and bound in the United States of America

10 9 8 7 6 5 4 3 2 1

Contents

Contents

PART THREE
After the Decade of Time

Acknowledgments

SANDFORD FLEMING'S feelings about the civil engineering profession—the so-called tunnelers, levelers, bridge-builders, and track-layers—were expressed in an 1876 speech: "It is the business of their life to make smooth the path on which others are to tread." On a later occasion, he reflected on the near-tragic art of engineering. If the civil engineer has done his job well, all traces of it disappear and others take the pleasure, the profit, and the recognition.

The Fleming papers in the National Archives of Canada indeed bring pleasure and smooth the path. Each alphabetical division in the 145 boxes (ironically, they are not chronologically arranged) contains its share of juxtaposed surprises and leads, I hope, to some piquant moments in this text. Typed letters from world leaders in his honor-encrusted old age rub against drafts of condolences sent to an Indian guide on the death of his wife, forty years earlier. I am especially indebted to the staff of Hutchison House in Peterborough, Ontario, who sent me Fleming material the moment I first expressed interest, and allowed me to browse through their collections when I visited a few weeks later. Otherwise, I have cited, where appropriate, debts to the major scholars in the field, who, like Fleming, are always there to smooth the way: William Everdell, Michael O'Malley, Pierre Berton, Jacques Attali, Stephen Kern, Derek Howse,

viii   *Acknowledgments*

David Landes, Peter Gay, David Harvey, Wolfgang Schivelbusch, Arno Borst, Eviatar Zerubavel, James Burke, Walter Houghton, and to the hosts of other books and articles that informed them, or trail in their wake. My former colleagues at the University of Iowa Mitchell Ash, Ed Folsom, Shelley Berc, and Garrett Stewart helped inform my reading early in the process and deserve a salute from two time zones away.

CLARK BLAISE
San Francisco
March 2000

Foreword

THE GAUGE AGE

NATURALISTS AND PSYCHOLOGISTS believe that of all forms of animal life, only man possesses a sense of time. It may be our defining characteristic. The historian of ideas Daniel J. Boorstin devotes the first three chapters of *The Discoverers* to time and the history of its measurement, because nothing is more fundamental to our nature than the observation of time and the struggle to measure it accurately. Without an agreed-upon standard of time, we cannot mark or measure change. There can be no innovation, no discovery. Like Robinson Crusoe notching a stick, or prisoners in the Gulag scratching a line for every day of their confinement, we are embedded in time. Even when we leave society behind, our very sanity depends on periodicities. What day is it? How long have I been here? The way we know time today has a great deal to do with the creation of standard time, and the man this book, in part, celebrates.

His name fades with each new generation, although plaques and memorials abound. A college and a few secondary schools are named for him, but fifty or sixty years ago, Sir Sandford Fleming would have won the possibly self-ironizing title of "outstanding Canadian of the nineteenth century." Born in Scotland, in the manufacturing town of Kirkcaldy, in 1827, the son of a local contractor, he received his six years of formal education in the town (burgh) school, and then apprenticed himself another six years to the local land surveyor, John Sang. In 1845, at the age of eighteen, he and his older brother sailed for Canada. A cousin presented him at the docks with a silver sovereign. His father entrusted to him a valuable watch with a built-in sundial, an emblem of the time system he would eventually overthrow. The ticket to his future on the sailing ship *Brilliant* cost the sizable sum of £4, for which he and his brother were guaranteed a daily quantity of drinkable water and basic uncooked rations.

In the towns and cities of Scotland, horse-drawn omnibuses made regular stops, their painted sides announced their final destinations. "Glasgow Docks," it might say, "with train connections to Liverpool" to meet all sailings. The young men whose wooden trunks contained meager treasures of books and professional instruments, the tins of flour and tea, and the bedding, were headed for Canada, South Africa, New Zealand, and Australia. Some chose the more alien challenge of the United States. Emigration was the inescapable destiny of the bright and enterprising Scotch, as they called themselves: populate the Empire, build the machines, run the engines, make their fortunes. The lessons of their straitened childhood and the strictures of the Presbyterian Church kept them sober and responsible for the rest of their lives.

They were the steamfitters, the boilermakers, the gauge-readers, the engineers of the world, proud of their hardiness and frugality, quick to grasp the mechanical advantage. Victorian pop psychology assigned various aptitudes to distinct "races," and the Scotch were thought to have an uncanny affinity for technology. (The stereotype carried forth to our own fantasies in which "Scotty" worked his mechanical magic on the starship *Enterprise*.) Kirkcaldy, on the north shore of the Firth of Forth, across the water from Edinburgh, was also the birthplace of Adam Smith. Thomas Carlyle had served as master in the same burgh school a decade before Fleming's matriculation. Linoleum was invented and manufactured in Kirkcaldy. There's probably not a significant settlement in Scotland that could not provide a comparable list of famous sons and their useful inventions. Fleming's later friend Andrew Carnegie, one of the America-bound, hailed from the nearby Fifeshire town of Dunfermline. Andrew Cunard, the shipping magnate, and the parents of James J. Hill, founder of the Great Northern Railroad, the first two prime ministers of the Canadian nation, Sir John A. Macdonald and Alexander Mackenzie, and untold bankers and businessmen, who collectively established Canada as Britain's leading colony among presumed equals, had all made the same passage and adjustment from Scotland to Canada.

What they carried with them was their faith, their confidence, and their "genius for hard work." They remembered Scotland fondly, returned often, and contributed materially to its survival. They got on well with Americans, and many, of course, like Carnegie or Alexander Graham Bell or J. J. Hill, are associated almost entirely with their American successes.

There's a subtle parallel to be traced between Scotland and Canada, two non-countries by the standards of Victorian diplomacy, unrecognized, even threatened, by their powerful southern neighbors. The exuberant reticence of the Scotch—sober, hardworking, calculating to the last penny—was particularly appreciated in the underpopulated void of autonomous colonies called British North America before the 1867 British North America Act that created Canada. Popular opinions of the Scotch were quite a bit more flattering than the general view of their fellow Canadians, the Irish and French. Doubtless, the Victorian mind distinguished them for their sturdy Protestantism. But the Scotch, like the country they had left and the new territory they were building, were negotiating a very tight passage between proud survival and overt surrender. They were "emigrants," not immigrants. They had known poverty in their homeland, but

overnight, it seemed, had been transformed into hardy trans-
plants in Canada, the United States, or in England itself. Fleming's life is one long demonstration of competing loyalties to
Canada, to Scotland, and to the idea of the British Empire. As
a prime example of the successful emigrant, he nevertheless
lamented on return visits to Scotland the loss of his distinctive
accent, and even his ear for the purer strains of the "north of
Tweed" dialect. Only in Kirkcaldy was he taken for a native.

The Fleming brothers nearly died on that forty-four-day passage in 1845. On one fearful night in the midst of a North Atlantic
gale, Sandford took readings of wind speed and direction, calculated the ship's heading and tonnage, and determined that they
might not survive until morning. He inscribed that sober assess-
ment, adding a declaration of faith in God and a profession of fil-
ial gratitude, bottled the note, and threw it overboard. Naturally,
his life being one long monument to industry and good fortune,
the bottle was picked up on a North Devon beach and delivered
to his parents not many months after he'd settled in Peterborough, his first Canadian home. He kept that letter in the top
drawer of his desk the rest of his life, along with the never-spent
silver sovereign his cousin gave him at the Glasgow docks. In Peterborough he found no work, only discouragement. He traveled
to other Ontario towns, taking surveys and striking town maps
off his own lithography stones and selling them. His notebooks
detail every sale, every expense. Half of his earnings were sent
back to the family. Within three years, his parents joined him in
Canada, settling on a farm their son had managed to purchase.

And one should not ignore the lesson of his teacher, the land
surveyor he had left behind in Kirkcaldy. John Sang and his sons
remained an important force in Fleming's life. Kirkcaldians remembered Sang as "a practical and mechanical genius" (such
praise fairly defines the social ideal of Victorian Britain), with a
particular aptitude for turning out engineering students. His

headquarters were more a technical college than a surveyor's office. He invented an instrument—an elaborate gauge, a converter with a readout—for automatically measuring acreage from a map, by tracing the perimeter of the area in question.

Sang's invention—and Fleming's more abstract inventions for world standard time—were rooted in the Victorian era's great facility and fascination with gauges, which are more than simple needles on a graduated dial. Steam power, the primary energy source of Fleming's day, was inherently dangerous and demanded constant attention, the gauge being the only practical way of monitoring internal heat and pressure.

The gauge is an intricate conversion device, a kind of translator between unspoken languages. Scales, thermometers, watches, fuel pumps and all their myriad applications are gauges, instantly declaring the equivalence of disparate and invisible events. "You've lost weight . . . you have a fever . . . you're running low . . . you've got ten minutes . . ." Like crude computers, they monitor one set of operations and convert it to a different data flow—to time, to temperature, to volume, to cost, to profit, to depletion. Any new invention with the hope of catching on in the nineteenth-century marketplace fairly bristled with valves, and glittered with elaborate brass displays, like the illustrations accompanying Jules Verne's undersea and lunar adventures or, much later, the time machine of H. G. Wells. They allowed Victorians to read, and to trust what they couldn't see. The Steam Age was the Age of the Gauge, a technology of convertibilities, which also had a strong influence on the standard time movement.

Sang's story would end unhappily. In later years, he counted himself a "coward" for not going to Canada, where Fleming had offered to set him up, or to Wisconsin, where other Kirkcaldians had settled. From two lives that began so much alike, two very different destinies can be traced: those who left and those who stayed. John Sang and his sons lost everything they'd built, they sold their instruments and declared bankruptcy, ending up working for others, copiers and deed certifiers, in a Belfast office.

By contrast, Fleming's life is a story of ever-growing success. Five years after arriving in Peterborough, and then Toronto, he had made his mark as a surveyor and lithographer. He took the first soundings of Toronto harbor and struck the first map of the city's streets, including the harbor and beaches, then wrote papers on the geologic history of Lake Ontario and its successive prehistoric ledges. He lithographed other surveys of Ontario towns, selling copies both remarkably accurate and artful. He was a fine amateur artist and often illustrated his own work. From surveying he made the logical leap to civil engineering, built railroads, designed and engraved the first Canadian postage stamp (the "Beaver," valued for its industry and engineering skills, not its pelt), and founded the Canadian Institute (which grew into the Royal Society of Canada), where he would deliver most of his scientific addresses, including the classic papers on standard time. He wrote a dozen books, served thirty-five years as titular Chancellor of Queen's University in Kingston, Ontario (in preparation for which he cited his six years of training in a Scottish town school), devised and facilitated world standard time and, finally, the world-circling sub-Pacific cable, which earned him his knighthood in 1897.

The *Montreal Gazette* characterized him, late in life, as "a man not happy without some great reform." Among these were: the Presbyterian prayerbook; metric/imperial unification (urging the French, with predictable results, to raise the meter's length to forty inches from 39.37, so that the systems could be made fully convertible); proportional parliamentary representation; and stock-market accountability. With a membership in over seventy international societies, he was for half a century Canada's voice on the world stage. For all of that, at the moment of his greatest success, the Prime Meridian Conference, which he had orchestrated, he suffered a failure, a bitter failure, partly of his own making.

THE SUBJECT of time holds a universal fascination, but this particular inquiry is also inspired by the appeal of one man and his age. Sandford Fleming put me in touch not only with the richness of the Victorians, our direct ancestors, but also with the country of my parents and of nearly half my life. Today, the differences between Canada and the United States are small indeed, but such was not the case a century and a quarter ago. My niche in the writing world has been to mark and measure those disappearing distinctions.

I turned sixty years old while writing this book, but I am still the avid child listening to his mother's stories of Manitoba and Saskatchewan, and still uncovering the untold tales of his father's Quebec. As I sifted through the boxes of Fleming letters in the National Archives, I could hear their long-dead voices afresh, and kept repeating the mantra that informs so much of this book: *it's about time.* It's all about time.

Part One

A (VERY) BRIEF

HISTORY OF TIME

1

The Discovery of Time

AN OBVIOUS question demands to be answered from the out-set: Can *anyone* have a definition of time? Time is invisible and indescribable, endlessly fascinating and universally compelling. Time is everywhere; thus nowhere. It animates the world, yet nothing survives it. We can only guess how it started, or when it will end. It is our intimate assassin. One thing it lacks, however, except in Greek myth, is a compelling narrative.

Natural time—the time of the gods, the sun and the moon—starts in a savage, glorious myth and ends on an Irish railway platform in 1876, when Sandford Fleming missed his train. Orig-inally, Time was embodied in a god, Uranus. He ruled over an immutable world. His children were the seven visible planets. Acting on a prophecy that his life was in danger from one of them, Uranus did the natural thing and slaughtered them all. Their mother, his sister Gaia, was able to hide one son, Kronos. Kronos, upon maturity, did the natural thing—castrated and killed his father. He married his sister, Rhea. When he learned of a plot against him, he cannibalized his children, but for Zeus, whose sleeping body Rhea had replaced with a stone. Zeus, of course, would castrate and kill his father.

Time is a bloodthirsty savage. None of us gets out alive, re-gardless of piety, decency, beauty, or innocence. But Zeus, at least, made it tolerable by setting the clock of mortality and mutability. We die, but we are replaced. Our children do supplant us; and they bury us. They can't admit it, but *they want their parents dead.* And parents can't admit it, but *they want their children forever helpless and dependent.* So long as they remain babies, we stay in our virile prime. Their maturing is our death. Mutability saves us from unthinkable violence, at the cost of our own life. I can't imagine a more ethically charged dilemma.

The powers of Time were scattered. Different gods attended to prophecy, history, fate, and dreams. Priestly castes learned the natural periodicities of the days, months, and years and determined rituals and sacrifices required for harvests, protection from floods, and return of the rains. Natural time is cyclical, a closed system, not admitting to change. Gods of the natural world are mysterious, unknowable, and violent. Any variation in worship might—or might not—bring instant death.

The collapse of "natural" thinking was most sudden, and most dramatic, in England. It was in England that the Romantic embrace of nature reached doctrinal intensity, where rambles in the Lake District inspired poetry, where the great and permanent forms of nature were invoked as guides in time of crisis and despair. One thinks of nature's power not only to soothe but to inspire reveries of timelessness, as in Keats's "Ode on a Grecian Urn" (1818). But England soon embraced the Industrial Revolution with even deeper fervor, so that within little more than a generation, the nation had been transformed into a virtual laboratory for creative destruction. Thirty years after Keats's ode, in *The Communist Manifesto,* time became cheaper than sand, not dearer than gold, a servant, not a master. It could be leased back to an employer at a fair rate and for a set duration, or even confiscated by the proposed new state on the behalf of labor. The social structure and the political order were transformed, but not by Marx and Engels. The revolutionary agent was speed, the new velocity introduced by trains and the telegraph. If industrialism and rationality teach anything, it is that nothing is permanent, especially nothing found in nature. There is no "natural" law. Displaying gratitude for the gods' gift of time became less important than showing up punctually for a day's work and collecting a guaranteed wage at the end of the week. Standard time, which also arrived in Britain in 1848, is the ultimate expression of human control over the apparently random forces of nature.

WRITERS WHO find themselves fascinated by some aspect of time usually confess their inadequacy, or their confusion, by invoking St. Augustine's famous admission in his *Confessions*. It reads in essence: "I know what time is, but when I try to describe it, I cannot." That leaves quite an opening for anyone who would rise to the challenge. The English historian Simon Schama, in the opening of *Landscape and Memory,* speaks directly to the issue:

> For a small boy with his head in the past, Kipling's fantasy [*Puck of Pook's Hill*] was potent magic. Apparently, there were some places in England where, if you were a child (in this case Dan or Una), people who had stood on the same spot centuries before would suddenly and inexplicably materialize. With Puck's help you could time-travel by standing still. On Pook's Hill, lucky Dan and Una got to chat with Viking warriors, Roman centurions, Norman knights, and then went home for tea.

The American physicist George Smoot, combining astrophysics with autobiography, begins his *Wrinkles in Time* on a simpler note: "There is something about looking at the night sky that makes a person wonder." A few sentences later, he brings that childhood wonder up to date:

> I could discover not only new things, like ponds and tadpoles, but I could also find out what caused things to happen, how

they happened, and how things fit together. For me it was like walking into a dark museum and turning on a light. There were incredible treasures to behold.

Fiction writers attracted to time can only envy historians and astronomers. Time, after all, is their raw material. Novelists are no less wonder-struck, no less time-besotted, and no less driven to fit things together, but their tool is the story, the actors and plot. Their approach lies closer to the way of sociology and psychiatry (or perhaps forensic science), stopping time, fragmenting it, backing it up, moving it forward, examining the pieces. Time lacks that narrative base, it is so nebulous that it might evade definition all together, by anyone. "Time is like Oakland," the sociologist Murray Davis once said, echoing Gertrude Stein, "there's no *then* there."

First of all, time comes in two distinct varieties: the untamed, mysterious Time, born with the big bang itself, and civil, obedient standard time, as in "What time is it?" or "How long has this been going on?" It's not clear that the same word even applies to both, or what the nature of their relationship, if any, might be. Perhaps time should have two names, like "horse" and "equus," the one to stand for hardworking, domesticated time, that which we control and can describe—the calendars, clocks, minutes and hours of the civil day—and the other for the untamed and unnamable, that which nature has not yet released.

The cesium-ion atomic clock is so accurate that it "loses" only one second every ten thousand years, and even that exact standard is open to further precision. It divides each second into more than twenty billion pulses. But what exactly is it dividing, what is it measuring, what is a second, what is a minute? And if we "lose" it, where does it go? When basketball games are won or lost in the final seconds, or when downhill ski races and Olympic dashes are decided by tenths, hundredths, thousandths, or ten-thousandths of a second, are we honoring accuracy or exposing the arbitrary nature of measurement, the meta-measuring of measurement itself? It seems apparent that some contests are not won or lost in head-to-head competition, but in the anachronism of relying on a starter's pistol, our inability to mark a true beginning—or, in terms of this book, our failure to fix a proper prime meridian. A smart lawyer could argue that a runner lined up in an outer lane, twenty yards away from the starter's pistol, hears it a significant thousandth of a second later than a runner five or six lanes closer.

The irony is inescapable. Ever-finer precision creates everwidening ambiguity. The nineteenth century's faith in rationality led supremely confident rationalists in anthropology, sociology, and psychology to study the presumptions behind civilization, or reason itself. Only confident rationalists could explore the irrational, but once they got there, what they discovered undermined the confidence that had got them there in the first place. Enthusiastic evolutionists, as most late-Victorian scientists were, believed they'd been given a key to understanding far more than the origin of species. They saw evolution as applying to history, society, economics, to God, the cosmos, language and logic and the mind itself. Thomas Henry Huxley, the great apostle of Victorian science, believed, in 1887, that applied evolutionary theory would deliver a unified explanation of everything—biology, physics, chemistry, and religion. In 1879, Leslie Stephen, introducing the essays and lectures of his polymathic classmate William Clifford, who had died tragically young of tuberculosis that year, recalled their undergraduate enthusiasm for rationalism in all fields:

> Clifford was not content with merely giving his assent to the doctrine of evolution; he seized on it as a living spring of action, a principle to be worked out, practised upon, used to win victories over nature, and to put new vigour into speculation. Natural Selection was to be the master key of the uni-

verse; we expected it to solve all riddles and reconcile all contradictions. Among other things it was to give us a new system
of ethics, combining the exactness of the utilitarian with the
poetical ideals of the transcendentalist.

Two years after writing this, a formidable presence appeared
in Leslie Stephen's life, his daughter Virginia. She would grow
up just as High Victorian certainties were yielding to doubt. Her
generation would devote their creative lives to the refutation of
nearly all the comforting, steady-state theories of consciousness
they'd ingested in their privileged, progressive childhoods. The
scientific and material advances that gave leaders of the Victorian
establishment, like Leslie Stephen, their faith in reason became a
pompous culture of confidence to Edwardian progressives like
H. G. Wells, and a target of ridicule for Oscar Wilde. A generation later, Lytton Strachey, D. H. Lawrence, and Leslie Stephen's
daughter Virginia Woolf saw their complacency as a pathology of
posturing.

The thing we call time, in other words, is very difficult to dis-
entwine from the ways we measure it, from language, social convention, or the internal clock of our DNA. It cannot be described
in terms outside of itself. It is, as St. Augustine discovered, a tau-
tology. Many things are like time, but time is only like itself. What
can be described, however, is the history of standard time, clock-
and-calendar time, the man-made system of time-reckoning. The
great achievement of standardization in the nineteenth century,
culminating with the Prime Meridian Conference in 1884, was
to rationalize "real time" over thousand-mile (or fifteen-degree)
zones, and to give it a starting line, Greenwich, agreed to by all.
Thanks to standardization, we had the man-made tools to calcu-
late that New York's four o'clock was simultaneously Chicago's
three o'clock, London's nine o'clock or Sydney's . . . whatever.
(At least, we know how to figure it out.)

It is useful, of course, to know what the "real" time is in other parts of the world when we make telephone calls and run the risk of waking up real people from real sleep, though it hardly matters to e-mailers or stock-traders. Standard time as we've inherited it is the final great achievement of Victorian rationality; unreformed, it is vulnerable to the same forces that swept away other golden keys to understanding. (I can't imagine that twenty or thirty years from now we will still be computing time on a foundation that we've inherited, unchanged, from the age of steam.) Adjusting to new time will be like learning a new language, perhaps very much like learning a new language, if we're hard-wired with space-time coordinates, as Immanuel Kant originally proposed, or as some modern followers of Noam Chomsky might endorse.

TIME HAS only one visible analogue, and that is space. For Fleming, leader of the standard-time movement in the late nineteenth century and a trained surveyor, time and longitude were interchangeable. He even devised elaborate new watch-faces to prove his point. A glance at the outer wheel of longitudinal letters would give the time, and the inner wheel of clock numbers would disclose the longitude.

In 1860, when he was then a thirty-three-year-old surveyor and civil engineer, the University of Toronto selected him as external examiner for the first-year course in Surveying and Geodesy. John Sang would have been proud. The test he set bears a close resemblance to his own apprenticeship as a teenager in Scotland:

1. Give a general description of a theodolite, its construction, and the uses of its essential parts.
2. What is understood by the *line of collimation,* and how is the *error of collimation* detected and corrected?
3. Describe the construction and uses of an *optical square.*
4. Describe generally one or more methods of conducting a trigo-

nometrical survey and protracting the same, also the instruments employed in the field and office.

5. Explain the principle of the vernier.
6. State how the latitude of a place is ascertained.
7. What is understood by *magnetic variation,* as well as by the changes in the variation?
8. Give a description of one or more methods by which a true meridian may be determined, pointing out the comparative advantages of each method in practice.
9. Point out how the longitude is formed.
10. From the following bearings and distances, protract the figure, prove the accuracy of the bearings, correct the error, if any, and find the approximate area: [a six-sided figure was given].
11. A solid has two parallel ends 128 feet apart; the area of one end is 450 square feet; that of the other 270 square feet; find the number of cubic yards it contains by the prismoidal formula, and by any other method.

In many ways, Fleming's surveying test of 1860, conceived to judge ability in the measurement of space, applies equally well to time. A vernier (named for its French inventor), incidentally, is a kind of gauge, analogous to the second hand on a watch, that can be attached to a larger measuring instrument in order to provide an instant readout of more precise divisions of distance. The theodolite, then as now, is the basic instrument of surveying and civil engineering. Mounted on a tripod and precisely leveled, it measures vertical and horizontal angles. When the "lines of collimation" are connected, uneven surfaces are converted into a precisely rendered grid. In 1860, determining one's precise geographical location was a thoroughly rational profession. Temporal positioning, by contrast, was still arrived at by solar approximation. Before the decade had ended, in 1869, the first tentative proposal for linking time and longitude—that is, for rationalizing the dimensions of time and space—would be launched.

The surveyors' instruments have their great and small applications, from establishing the earth's longitudes and building railways, to determining property lines. They have analogies to timekeeping. Verniers and theodolites date from the sixteenth century, and both were in Fleming's trunk carried from Scotland. When John Sang and his sons were forced to sell their instruments in a bankruptcy proceeding, they were giving up the tools of their identity.

O N T H E quantum level, as we now say, 140 years later, time and space are indistinguishable, as they are in relativity, as they presumably were before the big bang, and as they are over the "event horizon" of the black hole. Time and space are, in some ways, identities; they can be expressed in mutual terms—"a day's ride," "a ten-minute walk." A student of mine in Montreal, fluent in English but foreign to some of its idioms, once listened to a weather forecast of "five quick inches" of snow on an English-language radio station and asked, panicked, "How big is a 'quick inch'?" Space, like time, has been measured civilly, legally, astronomically, and politically.

The meter, which is one ten-millionth of a quarter-arc of the earth from equator to pole as measured by the French two centuries ago, and offered to the world as the objective standard for all measurement, can be alternatively defined as the distance light travels in .000000003335640952 of a second. But the fact remains: we have measured time with extraordinary precision, and we still haven't seen it and can't say what it is. And of course, the meter is not "objective" at all; it's merely French. The Germans measured the same quarter-arc in the 1880s and came up with a different figure, a German meter. Today's measurements by laser from orbit revise it further: an American meter. We are post-Heisenberg; we know that we can't escape our subjectivity. We're postmodern; we can't ignore our cultural bias.

We know that time and space are being "created" (like our

lives, like the expanding universe, like the future) and that those times and places will be rich and deep—but that we'll never see them. It is an insult to our intellectual vanity, our Faulknerian and Keatsian and Trekkie souls that we'll never leap ahead to the future we imagine, nor ever walk the streets of an historic Else-when. We yearn to. The dream of stopping time and shrinking space is a marker of our humanity. Our dream is of universal, eternal, and instant communication. Our minds soar with instant connection, but our feet are stuck in temporal boots. At the outer rim of our soaring ambition, we are still confined by time, the unbreakable barrier of the speed of light, which is another way of saying the speed of time. Some day perhaps we'll make those journeys by virtual reality.

O F A L L the inventions of the Industrial Age, standard time has endured, virtually unchanged, the longest. We can say where and when standard time for the world started: Bandoran, Ireland, in June 1876. We can say who created it: Sandford Fleming. And why? He missed a train. His fault? No, a misprint. Why a mis-print? Because, unthinkingly, we double-count the hours of the day and *A.M.* can be printed as *P.M.* and we're too lazy to count above twelve. That moment of frustration in 1876 became an infinitesimal pinhole through which history and culture were projected.

Arguably, standard time has exercised the deepest influence on everything to come afterwards. The various manifestations of world standard time—the Greenwich prime meridian, the inter-national date line, the unification of the various professional "days," the twenty-four-hour clock, the counting of longitudes west and east from Greenwich, the definition of the "universal day," and, by implication, the most important of all, the twenty-four time zones—all came into existence by diplomatic and sci-entific agreement at the close of the Prime Meridian Conference

held in Washington, D.C., October 1–22, 1884. The conference
was officially called by President Chester A. Arthur and opened
by his redoubtable secretary of state, Frederick Frelinghuysen,
but the force behind it, its orchestrator, was Sandford Fleming.

The standard biographies of Chester Alan Arthur, deservedly
one of America's least celebrated presidents, do not even mention the single great achievement of his term, his role as the host
and organizer of the Prime Meridian Conference. It is one of the
smaller ironies of American history that a sweeping, international
event like the settling of the prime meridian and the protocols of
world standard time, which brought a distinguished gathering
of leading astronomers and diplomats from the world's twenty-
six independent countries to take part in one of America's earliest
assertions of diplomatic influence on the world stage, should
have occurred on Arthur's otherwise mendacious watch. A month
after the Meridian Conference, as related in Adam Hochschild's
King Leopold's Ghost, his friends were conniving with the agents
of the imperial powers carving up, all too literally, the African
continent.

In one way, however, Chester Arthur was perfectly placed by
history to understand the issues of standard time. He worked
well with entrenched power, particularly the railroad establishment, and was disinclined to challenge it. Prior to his assuming
high national office, the spoils system had installed him as customs chief of New York Harbor. The same Republican-party
machinery placed him on the 1880 ticket with James Garfield.
The presidency was then handed to him by an assassin's bullet.
With his gracious manners, charm, honorable Civil War and
antislavery record, his basic decency—and those impressive ax-
blade muttonchops—he embodied the plush, don't-rock-the-
boat certainties of the Gilded Age. He knew railroads, counting
many a rail baron among his friends and patrons, and loved to
travel in luxury accommodations. It could be said with a certain

degree of admiration that he was one of the least ambitious men ever to hold the office. Yet if any American president, apart from polymaths like Thomas Jefferson and John Quincy Adams, or perhaps Jimmy Carter, understood the value of standardization in weights and measures and the role of railroads in forcing change, it might well have been Chester Alan Arthur.

Time and Democracy

THE WIDER, extended narrative of standard time started before that day in Bandoran, Ireland, in 1876, and did not end in the Washington conference in 1884. It did not even start with the involvement of Sandford Fleming. Fleming was the catalyst, the man with a vision at the tipping point, when conventional responses to a growing problem were no longer sustainable.

Standard time culminated the long march of reason that had begun with the Renaissance. It coincided with the harnessing of a power source that transcended the mechanical limits of human and animal muscle. Perhaps the first great technological advance leading to the Industrial Age began with nothing grander than a teakettle and a curious, anonymous child who observed its top rattling and lifting under the build-up of steam. If so little water so inefficiently channeled could do *that,* what might more steam, aimed and tightly directed, do? What others saw as a noisy irritation, someone saw as a cheap, easily imitated power-source. Perhaps he grew up to be Giovanni Branca, who in 1628 created a crude steam turbine. Or perhaps he had the misfortune to have been born in France as Solomon of Caus and be confined to an asylum by Cardinal Richelieu for theorizing that the power of steam could out-perform man and beast. In the 1690s steam was powering Thomas Savery's inefficient vacuum "fountain" pump, but a decade later Thomas Newcomen's refinement of the same design was pumping water out of England's deepest mineshafts.

Unless the ever-deeper mines of Newcastle could be pumped dry, England faced a serious crisis in providing sufficient coal for the open-hearth ovens of its growing iron industry. It took another half-century for the expansive power of steam, and—with the addition of an exterior condenser unit—the contractive force of the vacuum, to be combined in a single effective energy source, James Watt and Matthew Boulton's reciprocating steam engine (1769). It is the basic invention from which all rotary movement (thanks to Watt's further refinements), including the railway locomotive, takes off.

But steam power had to be wedded to rails before the story of standard time could truly begin. Learning to take coals *from* Newcastle underlay the eventual development of standard time. In the Tyneside coalfields in 1630, young Master Beaumont introduced a system of wooden tracks that permitted a single horse to haul upwards of sixty bushels at a time. Any increase in load capacity, or in efficiency, is indirectly a temporal calculation. Cast-iron rails were introduced to the collieries in 1767, overcoming the commonsense assumption that smooth wheels on even smoother tracks would "naturally" slip and never grip. Another colliery engineer, Master Jessop by name, added the stabilizing inner flange to the iron wheels. The first "iron road," the Surrey Iron Railway, was chartered in 1801.

IN A WORLD totally dependent on horsepower and sailing ships, time and distance were tangible barriers to the exchange of information. There were no technical expositions, no annual learned conferences. It wasn't easy, or even possible, to transport designs or working models from site to site. It was often a matter of luck that accounted for the right entrepreneur meeting the proper engineer at the moment of inspiration. One of those fortunate places for two hundred years happened to be the coalfields and mineshafts of the Tyne River valley, near Newcastle, where necessity and inventiveness and a certain amount of what we'd call today venture capital all came together. What those two centuries of slow progress and intermittent discovery demonstrate, even in the same confined geographic area of Newcastle, is the difficulty of generating synergy in a world of unawakened time.

With the introduction of steam energy, first as an adjunct to horsepower and eventually as its replacement, businessmen were able to project new models of cost and production, new dreams of higher capacity, lower investment, and faster delivery over wider markets. Rule-of-thumb calculation at the beginning of the nineteenth century, for example, placed the annual cost of maintaining a horse at four times the wages paid to an unskilled laborer. Reducing dependence on the horse was obviously a definitive moment in social and economic history. New business equations began to undermine every "natural"—that is, commonsense or inherited—calculation of energy expended to profit extracted. And if one asks where is time in all that coalfield, underground, horse-drawn activity, the short answer is "mechanical advantage."

Elapsed time is equal to distance divided by the rate of speed. The rate is affected by increases in power. As power increases, so does speed; time diminishes and so does perceived distance. It was the slow increase in speed and power—the fusion of rails and steam—that undermined the standards of horse- and sail power and, eventually, the sun itself in measuring time. Gradually, all those new ideas and new applications, moving in the same direction but at varying speeds, created a new comprehension of time and space. And so it took two centuries of steady incremental invention to bring the reciprocating steam engine, in the form of the locomotive, and the iron rail together. Once that happened, the pace of change increased geometrically.

The early Industrial Age, which closed the Romantic era (James Watt and John Keats both died in 1819), had challenged, or had at least redefined, the Romantic assumption that life was a contest between "mechanical" and "organic" sources of inspiration. The *Quarterly Review* in 1825 had laid down the challenge, boldly but myopically: "What can be more palpably absurd and ridiculous than the prospect held out of locomotives travelling twice as fast as stagecoaches!"

The answer came just four years later, on October 6, 1829, when Stephenson's "Rocket" won a competition (and a prize of £500) for drawing a twenty-ton load at an average speed of ten miles an hour over a distance of seventy miles in one day. That October morning in Rainhill, England, marked the big bang of the temporal revolution. Aldous Huxley, in his 1936 essay "Time and the Machine," stated it cleanly: "In inventing the locomotive, Watt and Stephenson were part inventors of time." Less than twenty years later, Britain had become so transformed by the railway that it united itself temporally under the time standard of the Royal Observatory.

From the 1830s onward, the rate of travel on land and water began to increase geometrically: fourfold, tenfold, a hundredfold, a rate that the cultural historian William Everdell calls, in *The First Moderns,* "a change in the rate of change." A year before the Stephenson "Rocket," in 1828, Sir John Herschel, Britain's great astronomer, proposed the first (and very technical) revision of astronomical time-reckoning. It would not have affected daily life in any way, but in essence, and by independent reasoning, Herschel had responded to the first industrial probing of the time-space continuum. Eight years later, Thomas Arnold, father of Matthew, witnessed the first train passing through the Rugby countryside and noted in his diary, "Feudality is gone forever." Institutions rooted in an ancient time-space continuum cannot survive the effects of geometrical extensions of either term in the equation.

Standard time is the unexpressed operating system of all interdependent technologies. It can be said that the adoption of standard time for the world was as necessary for commercial advancement as the invention of the elevator was for modern urban development. Britain's nearly forty-year lead on the rest of the world, temporally and industrially speaking, began at the moment when standard time was adopted. The first decade of standard time in Britain, the 1850s, was Britain's shining moment.

IN THE Victorian era, the ancient conflict between faith and science, the organic and the mechanical, was seen as a clash between modes of thought, not just sources of energy. The Victorians labeled them "natural" and "rational." Natural thought placed man in a created universe overseen by God and maintained by the fixed laws of nature. Time in the natural world was reckoned by the biblically sanctioned solar noon. Any challenge to revealed truth, any faith misplaced in a man-made creation, was condemned as "vanity." Science, technology, research, mechanical creations, and standard time were all vanities. Under the rational model, however, vanities (humans worshiping their own creations) replaced the natural God. Progress, not salvation, was the goal of man and societies. Standard time served most of the functions of God; it set the standards of trade and commerce, of justice and mercy.

Standard time, as defined by Western science and diplomacy, the time of treaties and contracts, overrode aboriginal time, Hindu and Buddhist, farmer and fisherman crack-of-dawn time, or setting-sun Muslim and Jewish time. The standard-time day begins at midnight in order to avoid the irregular sunrise and sunset of nature. Standard time is a god of predictability and precision, no longer the Grim Reaper, no longer the moral accountant of sloth and enterprise, but a mild Victorian gentleman. He shows up for work at Greenwich precisely at midnight, every midnight, for all eternity. He polishes the machinery, tightens valves, reads the gauges, and goes to sleep. He's more than a little bit Protestant. He expects *you* to be accountable and to show up when you're supposed to, and feel a little guilty if you don't.

I N   1 8 3 4   an American by the name of Ross Winans invented the bogie, the independently suspended, four-wheel assembly mounted at the ends of each railroad carriage. From that moment on (to choose one of many), European and American history parted ways. The most influential application (trains) of the nineteenth century's dominant technology (steam) separated, creating different concepts of travel and different designs for mass transport. Major differences between European and American civilization can be predicated on an apparently obscure technical design element.

The bogie, with its double-axle design, stabilized American carriages, which could then grow longer and heavier. The stability allowed American trains to trace curving trackbeds. American carriage interiors were of an open-bench design, like stadium bleachers (appalling examples of rampant populism to early European visitors) with a coal stove set in the center of the car. Passengers were free to roam about. European carriages employed a two-wheel, straight-axle design that limited the length and weight of their carriages and virtually eliminated the possibility of twists and turns or a contoured railbed. European railway carriages resembled a series of linked stagecoaches, each made up of six-seated independent compartments, served by its own door that opened to the station platform. There was no communicating corridor.

Since land values in North America were considerably lower than in Western Europe (even though labor costs were higher), there was less incentive for American engineers to plan faster and shorter routes, or to invest as heavily in bridges and tunnels. Even with higher labor costs, North American railroads could be built for about one-third the expense of their European counterparts. Thus, the slower American railroads followed the course of rivers, racing the steamboats and paralleling the canals, until the cheapest and most convenient bridging-site was reached. By 1865 George Pullman had begun constructing luxury cars, diners and sleepers, to take full advantage of the bogie and the extra hours an American passenger was likely to spend aboard. European engineers were forced to think faster, straighter, shorter, and damn the expense. In the case of railways, Europe became the home of speed, America of luxury.

And finally, the lessons of European history mandated barriers, not openness, between political borders. The compact dimensions of Europe, which could, or even *should,* have led to continental integration, did precisely the opposite. Deliberate impediments, such as discontinuities in track gauge and incompatible couplers, were introduced and jealously maintained. The same poisoned history kept the English Channel Tunnel on drawing boards for over a century. Mutual suspicion and prejudice also segregated Europe's Balkan and Iberian wings, preventing them from contributing to "Western" culture.

In the case of railways, it was feared that an integrated rail system would encourage Russia to invade Germany, or Germany to predate on France by the simple capture of one nation's rolling stock and riding it to victory. (It's not an unreasonable assumption; both standard time and the Autobahn came to Germany as logistical adjuncts to military mobilization. Field Marshal von Moltke, in urging the adoption of a single time throughout Germany, where there had been five standards, saw it as the civilian equivalent of military time.) Spain did not standardize its track gauge to the rest of Europe until 1968—an experience that first-time visitors of my generation remember as a middle-of-the-night transfer at the border outposts of Port-Bou and Cerbère. Until 1911, when France finally adopted standard time, Europe was an unregulated temporal nightmare.

The relationship between railway convertibility and the standardization of time, as always, was close and mutually dependent. Many European countries, following the early examples of Britain, Sweden, and Switzerland, coordinated their national times to their own national observatories—a Greenwich prime, a Paris, Rome, Uppsala, Bern, Copenhagen, Cádiz, or Berlin prime—but blocked any kind of international standard time. It was only marginally less complicated for a resident of Aberdeen, say, in the middle of the nineteenth century, to know the time in Berlin or Warsaw than it had been two hundred years earlier. History had trained Europeans to view temporal and mechanical convertibility as threats to their national security.

The struggle for the adoption of standard time was philosophical as well as technical, and goes back to ideas that have flared and subsided throughout human history. Time in its many disguises is part of the great debate over the just derivation of power. Who "owns" time? That is, who holds the ultimate right to negotiate its value—the worker or the boss? The tenant or the lord? The merchant or the priest? Elected officials or an inherited elite? Why are some born slaves to time, and others released entirely from its constraints? In this sense, the Magna Carta was a temporal event; the American Constitution a great temporal document.

In ancient China, the historian David Landes writes in *Revolution in Time,* time was a resource and, being a form of value, was the exclusive property of the emperor. Common citizens were even barred by curfew from sharing the night. Each new Chinese emperor was permitted to reset the calendar to his liking and, doubtless, for relief from his creditors. The priestly and imperial monopoly of time certainly had its uses in any royal court. Thousands of workers could be mobilized for decades of uncompensated labor with magnificent results, whatever the social cost, and we'll never see such results again: the European cathedrals, Great Walls, Pyramids, and Taj Mahals. When time is not equally distributed throughout a society, travelers are turned into nomads, workers into slaves, criminals into lifers, squatters into settlers. It is the horror of the eternal moment. There are no connections, no schedules, no laws. When time is an inherited private property, nothing that reduces its value can be negotiated.

Landes relates an instructive historical anecdote concerning technical inertia in a natural-time world. When Europeans were scratching the earth with crooked sticks, Chinese were using iron plows. When Europeans were using steel plows pulled by tractors . . . Chinese were using iron plows. "Without shared time," he concludes, "there was no marketplace of ideas, no diffusion or exchange of knowledge, no continuing and growing pool of skills or information—hence a very uneven transmission of knowledge from one generation to the next."

Chinese peasants were using iron plows three thousand years after their invention simply because there was no reason for them not to. Social forces, artistic styles, dress, food, and religious practices, like objects in Newtonian physics, persist in their being until restrained or deflected. And it's not just China; it's the inevitable result of "natural" thinking. When all behavior and beliefs derive from a single, infallible source, anyone who would defy it, or alter it, is by definition mad, or a heretic. (We don't have to journey back a thousand years to sample the persistence of natural thinking. In 1999, in the United States, one state voted to exclude the teaching of evolution and the "big bang" in state schools on the basis that "no one was there to observe it." *No marketplace of ideas. A very imperfect transmission of ideas.*)

Any challenge to established periodicity was treated, at best, as a harmless oddity; at worst, a heresy. Elaborate clocks brought by European visitors to the Chinese court were viewed as toys, their donors patronized as clever children. In a world governed by natural rhythms, the aberrant or the innovative is doomed to rejection and isolation, like Galileo, or poor Solomon of Caus. Why work harder? Why improve tools or work conditions? Iron

plows are good enough. If it was good enough for my grandfather, it's good enough for me. The Chinese court exercised a
time monopoly, and over the centuries the culture suffered for it.

The ultimate time theft is slavery, to be permanently on another's time, never to rest (except by malingering), never to possess (except through charity or theft). Instructive, then, that the
most creative manipulation of time in American music was the
invention and the property of slaves and their descendants. In
classical European music, the percussionist watches the score
and waits for his entrance. The skinsman in jazz, by contrast, is
in constant communication with time. He is his own Greenwich,
setting the tempo, creating the score freshly with every performance. In *Don't the Moon Look Lonesome,* the novelist and jazz
scholar Stanley Crouch writes: "In jazz, the time doesn't just
pass the way it does as it's defined on a ticking clock or a metronome; it interprets the tempo through swing, propels you, sup-
ports you and it talks to you, comments on your own activity and
you talk with it." Time talks to you, as it did to Keats peering at
the Grecian urn, and to Whitman, and to van Gogh as he absorbed the lessons of Japanese woodcuts, and, much later, to
Faulkner and Woolf and Proust, and as it has to nearly any serious thinker of the past century and a half. Time talked to them.
Time was in the air.

Our sense of a decent civil society depends on the rule of law,
but just laws, in turn, derive from what Landes called "shared
time," the democratic apportionment of time. Wages, contracts
and patents, terms of office and weighted sentences, permis-
sions, warrants lapsing and renewing, rents, interest, schedules,
penalties, bonds maturing and loans falling due—in all this, civil
society recognizes the beneficial impermanence of political and
economic activity. But not *total* impermanence. It seeks to preserve other institutions, and to render them time-resistant, as in
the case of life appointments, of tenure, of tax-free status for
churches, schools, museums, and certain kinds of foundations.

Democracy recognizes individual change as part of a greater continuity; change guarantees stability. Tyrannies shelter their institutions with permanence and resist all change as a threat to their legitimacy.

If time did not come directly from God, it came from the tsar, and later from the Communist Party. Young Cleveland Abbe, one of the major players in the standard time movement, friend and collaborator of Fleming's, president of the American Metrological (measure-reform, not weather-forecasting) Society, and founder of the U.S. Weather Service, spent two postdoctoral years (1866–67) in Russia working under Otto Struve, director of the Pulkovo observatory near St. Petersburg. During those years, he had to instruct his mother to stop addressing her letters to him at "the National Observatory," because, he explained, Russians had no concept of "national" apart from their language. Everything else was owned, named after, or donated by the tsar, like an indulgent father to his children. He was working in the tsar's observatory, under the direction of the tsar's astronomer. To the very progressive-minded Abbe, the tsar's authority had infantilized his people, turning their protests into futile acts of petty vandalism, their celebrations into drunken brawls, their marriages into loveless quarrels.

While there, Abbe fell in love with Struve's youngest half sister, Ämalie, and planned to marry her, and even to stay in Russia, or take her back to her native Germany. But he had not taken into account another aspect of time, the full weight of German conservatism. His formal request for Ämalie's hand was rejected by Otto. It was her duty as youngest daughter, he dictated, to look after her stepmother; and that is precisely what she did, right up to the years preceding the First World War. If they had married, and if Abbe had stayed in Europe and headed a European observatory, standard time assuredly would not have evolved as it did.

Following the rejection, he left Russia within weeks and eventually, after a few years heading the Cincinnati Observatory, reshaped his very productive life from academic astronomy to weather forecasting. In his Cincinnati years he, too, published standard-time proposals for reforming North American time-reckoning, featuring time zones and a Greenwich prime. It was as a weatherman, however, not a railroad executive, that the need for standardization pressed most heavily on him. In his Washington office, he received hourly weather bulletins telegraphed from dozens of reporting stations, hundreds and even thousands of miles away. They all had to be translated into a single isotemp, "real time," in order to predict the direction and magnitude of storm fronts, then plotted on his maps of isobars and isotherms (dreamy, elliptical, transitory lines, so different from the inflexible bars of longitude) and then resubmitted to, or retranslated in different local times for, subscribers in hundreds of daily newspapers.

Nearly twenty years after his great Russian adventure, Cleveland Abbe was reunited in Washington, as one of four American delegates to the 1884 Prime Meridian Conference, with Ämalie's half brother, Charles de Struve, the Russian ambassador to the United States and chief Russian delegate to the conference. The elective affinities of the world's intellectual elites were no less on display in the 1880s than they are today.

MY PARTICULAR family history is hardly exceptional among immigrant groups, especially considering that my parents were the least exotic of all foreigners in the United States: Canadians. It's rather like a Scotsman claiming immigrant status in England. Nevertheless, the silence that exists between familiars, or nations, often bears close attention.

My paternal grandfather, Achille Blais, was born to tenant farmers in 1865, the son of two hundred years of tenant farmers, in the Beauce region south of Quebec City. When he was nineteen, in 1884 (the year, coincidentally, of the Prime Meridian

Conference in Washington), he left the land and became a day laborer (*journalier*) for a dollar a day in the sawmills of a newly established village named Lac-Mégantic. He knew wood; he built furniture, he was a carpenter. His eighteenth child, my father, Léo Roméo, was born in Lac-Mégantic in 1905. My eighteen aunts and uncles, only five of whom lived past the age of seven, were named with an eye to eternity: Homer, Ovid, Hector, Ulysses, Iphygenia, Athena . . . down to the two youngest, Roméo and Rolland.

It broke my heart, a few years ago, researching the parish records of old Lac-Mégantic for an earlier book. Baby Ulysse Blais, dead at three years, six months. Baby Athénée, dead at two. They were all victims of the "natural" world of diseases borne in the animal and human putrescence that pooled in the stagnant water (kept artificially flooded by the lumber mill in order to float their logs). My grandparents serenely buried them and carried on. Eighty-five percent of their children, as Faulkner put it, "manured the earth." Two or three times a year, my grandfather affixed his "X" in the parish birth and death records, just above the priest who was the lone literate in the village. This is the way it had always been, and the way it would have stayed, in a natural world. Something in Achille, however, made it change. Time was in the air. Perhaps time had something to do with it.

He walked over the border, only ten miles away, and hopped a train to Lewiston, Maine. He worked in a shoe factory, and brought his family over. Nothing vastly heroic like Sandford Fleming's sailing the ocean and bringing his parents over in just three years—Lewiston being just seventy-five miles away—but the social change was arguably more profound. A few years later, they moved on to the Mecca of all French-Canadians, Manchester, New Hampshire, where the five surviving children found work in the spinning mills. No schooling, of course. In his lifetime my grandfather passed from "Achilles," a classical god, to the name of his obituary notice in New Hampshire, "Archie," a comic-book hero, an immigrant fate. But by the time he died he was a contractor and home builder.

In his seventy years he recapitulated not only the history of his people and their economic evolution—serf to entrepreneur— but the history of time itself. Grandfather, born into illiteracy, brutal mortality and a medieval Catholic world, entered history despite lifelong illiteracy because time, in the course of his lifetime, lost its godly authority. That dollar a day as a day laborer in Lac-Mégantic, as humble a time contract as can be imagined, led my father to a succession of marriages (at least one of them, to an *anglaise*, lasted for about fifteen years)—to Florida and to Pittsburgh, more marriages, and back to Manchester, where his life ran out. In two generations from tenant farming to house building, to my father's prizefighting, lounge singing, liquor running, and finally, furniture selling. I was freed for college, marriage to an Indian woman, and this life outside of time altogether, a "temporal millionaire," in the words of the social psychologist Robert Levine in *The Geography of Time:* someone, as he describes it, always ready for a movie in the afternoon or six months in another country.

A N Y A D U L T alive in the 1870s and eighties, that generation of time-makers, could remember a childhood when nothing he now took for granted was in existence, or even contemplated— electric lights, the telegraph and telephone, photography, refrigeration, typewriters, the train, the steam liner, theories of evolution, the molecular theory of matter and the conservation of energy. So much change in so little time; we can't imagine the excitement and the anxiety Victorians felt. There were suddenly no borders, no "natural" limits. Regimes that depended on containment for self-perpetuation, like the Ottomans or the Romanovs, found themselves outflanked in every possible way.

The outermost rings of time reform include the lifespans of

the major figures in the movement. When Abbe and Fleming started school in the 1830s, steam locomotion barely existed, but feudalism did, and the last Romantics were still alive. They grew up with the inventions of the electric cable and the expansion of the railroad grid, they absorbed the lessons of Darwin, and lived through the daily wonders of Victorian science and technology. They watched gasoline replace coal, electricity replace steam. Most lived to see the first flight and motion pictures and to listen to wireless transmissions. They believed they had seen the end of human want and misery, and that the cycles of natural calamity had been broken by the application of reason. They believed passionately in international cooperation, and they had the model of the Prime Meridian Conference to celebrate. They died reading reports from the battlefields of World War I.

Fleming bridges two worlds. He was born in 1827, the year Samuel Taylor Coleridge died. When he emigrated to Canada in 1845, there were only fourteen miles of railroad in the entire country. He built thousands more. He died there seventy years later, at his daughter's home in Halifax, on a day in 1915 when Commonwealth troops were being slaughtered at Gallipoli, in the year of Frank Sinatra's birth.

3

What Times Is It?

He and his eighteenth-century, troglodytic Boston were suddenly cut apart—separated forever—in act if not in sentiment, by the opening of the Boston and Albany Railroad; the appearance of the first Cunard steamers in the bay, and the telegraphic messages which carried from Baltimore to Washington the news that Henry Clay and James K. Polk were nominated for the Presidency. This was in May, 1844; he was six years old; his new world was ready for use, and only fragments of the old met his eyes.

—HENRY ADAMS, *The Education of Henry Adams*

BEFORE THE WORLD was divided into twenty-four time zones, and before each day began at midnight on the Greenwich prime meridian, and before the international date line kept the calendar in balance, every settlement with a need to know the time kept its own official clock. Time was based on the solar noon, that shadowless, sundial moment when the sun appears to stand directly overhead. But the sun keeps moving (or more properly, the earth keeps revolving), twelve and a half miles a minute, along the most populated latitude in North America. Imagine towns like knots tied on an infinite string; every twelve and a half miles signaled a new solar minute. Every eleven hundred feet, to be even more maddeningly precise, created a new solar second. Put in today's terms, every town was its own Greenwich. Adjacent villages clung jealously to their particular time with all the ferocity of a threatened identity, accusing others of keeping false time.

From the setting of an accurate local solar noon, public clocks and personal pocket watches could easily plot the precise time, and a "time-ball"—whose direct descendant is saluted each New Year's Eve in Times Square—could be dropped from the tallest steeple, or bells might be set off in the firehouse, or a cannon fired in the harbor. The time was accurate, it was indisputably twelve o'clock noon, but—and here is the great dilemma that faced the world until the 1884 Prime Meridian Conference—*it was only noon for that community, inside those twelve miles.*

All across the continent, every twelve miles-plus along the same east–west latitude—New York to San Francisco, let us say—as the sun crossed new longitudinal meridians, new noons were created, new time-balls were lowered, new bells went off. Out in the Delaware Valley, sixty miles west of New York City, it was just 11:55 A.M. when the whistles were blowing in lower Manhattan. In fact, when the time-ball was falling in Manhattan, it was only 11:59 A.M. in Newark, New Jersey, just across the Hudson River.

So long as people remained in their towns and villages, or on their farms, and so long as sailing boats, riverboats, barges and horses, or oxcarts provided the means of transport, competing time standards really didn't matter. No one could go far enough or fast enough in a day, let alone in an hour, to run into confusion. Then, as now, people wanted the assurance of precision, but they were not in the habit of looking beyond their own horizon. As "magnetic" (electric) technology evolved during the nineteenth century, it became possible for telegraphy to provide exact time signals from the nearest observatory, directly to the time-ball. The first such "magnetically signaled" time-ball was installed in Toronto in 1842. In 1868, at the Naval Observatory in Washington, the instant signal was coupled to an electric motor, which permitted the automatic dropping of a precisely accurate time-ball. Automatic signaling was an advance in convenience, but it did not address the basic problem. There were still too many different times. Had railroads and the telegraph—in other words, *speed*—not disrupted the system of local time notation, rendering the sun too slow a metronome, the temporal revolution might well have been delayed until the next revolution, the spreading electrical power grid, demanded it.

(Not to race too far ahead, but it is instructive, I think, to consider a temporal dilemma we will be facing in the near future. The Internet and its related technologies are establishing a new time consciousness even while we observe nineteenth-century railroad time as our official standard. The debates of the nineteenth century have shown us that society, and maybe even the human psyche, do not easily adapt to competing time standards. I doubt that the twenty-first century will be any more successful.)

ESSENTIALLY, ANY official adherence to the solar noon involves two distinct and unavoidable flaws. One derives from the earth's rotation, the other from its tilt. The rotation, as we've already seen, spawns countless noons, and therefore countless midnights. The circumference of the earth, twenty-four thousand miles, contains nearly two thousand solar minutes and a theoretical infinity of legal "solar days" based on the mad precision of the solar minute and the solar second. New York's "day" started and ended five minutes before Philadelphia's, a minute before Newark's, twelve minutes after Boston's . . . and so on. Ninth Avenue in New York is a fraction of a second "faster" than its westerly neighbor Tenth Avenue; they could each claim a separate day. Where does it end? If the Oakland Bay Bridge had been built in the nineteenth century, the Oakland end would have been thirty seconds "faster" than San Francisco's. At least,

the earth's rotation is steady and predictable, and if one is prepared to live with the infuriating multiplicity it demands, it is
calculable.

The earth's tilt, however, introduces a wild card. Unless one
lives on the equator, each day's noon falls a few seconds or minutes earlier or later than the noon of the previous day. The sun
rises or sets at a different minute each day. (The more northerly
position of Europe exacerbates the daily differences even more
dramatically.) In a world run on natural rhythms, regulated by
priests or other appointed or hereditary interpreters, solar irregularities are merely the visible face of God. Sunrise (for Hindus)
and sunset (for Jews and Muslims), rather than the West's civil
midnight, whenever they occur, mark the beginning of the natural or religious day. Only in a world of contracts, wages, and
schedules are natural variables taken as irritants. At whose midnight, for example—yours or the company headquarters'—might
an insurance policy run out, or a new law take effect? A birth or a
death occurring near midnight on the last day of a month, or
year, might officially occur (that is, be entered on a registry) on a
different day, month, or year wherever such records were kept, a
few miles east or west of the original event.

Versions of the sundial method of time-reckoning had served
the Greeks, the Romans, the Jews, the Arabs, and doubtless the
Egyptians, the Mayans, the Chinese, and the Druids—at least on
cloudless days—but their methods, however ingenious, could
not satisfy the needs for both precision *and* predictability that
technologically sophisticated Europeans and Americans came to
demand. In other words, nature's time is fine for religious observation, but it's a rotten way to run a railroad. And running the
railroad, as the rail systems expanded and nineteenth-century
commerce grew increasingly dependent upon them, is what
standard time was all about.

The true problem with ancient methods of timekeeping was
not only the inevitable meteorological and seasonal complica-

tions, but the fact that a multiplicity of so-called local times meant that none of them could be readily converted. Town A could not know the times in Towns B, C, and D, but this inability had not been a problem until rapid travel between settlements became common, practical, and comfortable. Solar time converts nothing; it cannot function as a gauge. Having too many times, the traveling public was discovering, was worse than having no time at all. Official times multiplied like weeds, and nature offered only a nightmare of endless proliferation. At mid-century, there were 144 official times in North America. And no one, apparently, was there to rationalize it.

SANDFORD FLEMING, in an 1890 essay, "Our Old-Fogy Methods of Reckoning Time," a retrospective history of the standard-time movement and the struggle for its acceptance, would have taken strong exception to a term I've been using rather loosely—"local time." He found the very idea objectionable, in that it encouraged the common misconception that there were, indeed, many times. When I use that convenient but imprecise phrase, please be mindful of the following caveat:

> "Local time" is a familiar expression, but it is entirely incorrect. There is no such thing. The expression "local time" is based upon the theory that time changes with the longitude; that each meridian has a separate and distinct time. Let us follow this theory out. Take a hundred or a thousand different meridians. All meridians meet at two common points, one in each of the hemispheres, the poles, so that at each of these points we would have a hundred or a thousand different "local times." This only requires to be stated to establish the impossibility and absurdity of the theory that in nature there is a multiplicity of "local times." There is one time only, it is a reality with an infinite past, an infinite future, and continuity is its chief attribute. It may be likened to an endless chain,

each link inseparably connected with its fellow links, while the whole moves onward in unalterable order. Divisions of time, like links in an unbroken chain, follow one another consecutively; they have no separate contemporaneous existence; they continue portions of the one time. Time remains uninfluenced by matter, by space, or by distance. It is universal and essentially non-local. It is an absolute unity, the same throughout the entire universe, with the remarkable attribute that it can be measured with the nicest precision.

That warning, six years after the Prime Meridian Conference that had settled standard time for the world, is a fair introduction to Fleming's engineering mind. Time was a unity with many facets (at least twenty-four), and he strenuously opposed the notion that any one claimant, even Greenwich, possessed a special time. It is the reason he fought stubbornly for the adoption of a "universal day" (sometimes called "cosmic" or "terrestrial"), free of any geographical determinant, free also of the familiar twenty-four time zones.

Through force of circumstances [his essay continues] we are now obliged to take a comprehensive view of the entire globe in considering the question of time-reckoning. We should not confine our view to one limited horizon, to one country, or to one continent. The problem presented for solution to the people on both hemispheres is to secure a measurement of the one universal passage of time common to all, which shall be based on data so incontrovertible and on principles so sound as to obtain the acceptance of the generations which are to follow us.

Those comments sum up the difference between Fleming and the American pioneers of standard time. He worked on a universal scale, they on a national. The next time I use the words "local time," please supply them with quotation marks, preceded by "so-called," courtesy of Sandford Fleming.

IN THE LATE 1840s, Henry David Thoreau sought refuge from the quiet desperation of town life and society by withdrawing to Walden Pond. *Walden,* like many classic texts, including the concluding "The Dynamo and the Virgin" chapter in Henry Adams's *Education,* can be interpreted by a willful reader as a reflection on temporality as well as society, particularly on the changing time standards that were (even then) in the air. Armies of workers were beginning to be regulated by time, the mechanical clock. "Actually," he wrote, "the laboring man has not leisure for a true integrity day by day; he cannot afford to sustain the manliest relations to men; his labor would be depreciated in the market. He has no time to be anything but a machine."

He has no time: surely a new and disturbing development. Leo Marx, Thoreau's great exegete, states in *The Machine in the Garden* that the function of the clock is decisive in Thoreau's version of capitalism "because it links the industrial apparatus with consciousness. The laboring man becomes a machine in the sense that his life becomes more closely geared to an impersonal and seemingly autonomous system." Certainly that is true, but there is another focus. Thoreau was alert to frightful new creatures that lurked in the temporal wilderness of mid-century Western culture. Men without time, without integrity. Machine men, emasculated men. He sought solace, like his English Romantic forebears, in the great and permanent forms of nature, in classic texts and Eastern religions. He was caught between the rapid spread of the railroad, which was refashioning the world in its own image, and society's helplessness before it. The railroad knew no temporal boundaries; men must bow before its demands. Thoreau's anxiety stemmed at least in part from industry's assault on what we'd call the time-space continuum. Time was in the air.

It was the railroad that filled the night with the creak and rumble of traffic, that brought the timbering and ice-harvesting crews, that filled the winter sky with clouds of black smoke. No wonder he remarked, "We do not ride upon the railroad; it rides upon us." *Walden* was an assertion of individuality resisting industrialization, an American version (and near-total inversion) of Marx and Engels's chain-rattling *Communist Manifesto,* not so coincidentally researched in the same year. In that universally revolutionary year of 1848, while France imploded, Britain prospered. She became the first country in the world to standardize time across her entire territory to the time signal of the Royal Greenwich Observatory. (Irish time was set twenty minutes slower.) That year also saw the publication of *Dombey and Son,* Dickens's almost literal demonstration of a railway's riding upon the back of a distraught, psychologically ruined Mr. Dombey, in which he noted, "There was even railway time observed in clocks, as if the sun itself had given in." Dickens, however, saw the train's power and brute human achievement as potentially restorative, at least to those susceptible to its message, like Mr. Dombey.

Time, then, was beginning its long association with business and industry, with schedules, with commercial entombment, with depression and anxiety. When I think of time's draining of personality, the enduring enigma of Herman Melville's "Bartleby the Scrivener" (1853) comes to mind, the eponym a ghostly presence in a Wall Street legal office. A character of compelling and enduring mystery, Bartleby is an emblematic wraith of negativity, an anomaly caught between temporal standards, as though waiting to die, or to be born. Present in body, absent in soul, very much the representation of Thoreau's worker without the "leisure for integrity," or the "manliest relations to men."

"EVEN THE EARLIEST settlers knew about clocks and watches," writes the American historian Michael O'Malley in *Keeping Watch: A History of American Time,* "but they understood mechanical timepieces as mere representations or symbols of time, not as the embodiment of time itself." In the middle of the nineteenth century the nature of time was beginning to change. It was no longer the God-given moral accountant of human folly (as in the parable of the ant and grasshopper: "It's seven o'clock, have you done your chores?"), but a new player in the sober world of commerce, punctuality, and reliability ("Oh, God, I've got ten minutes to get to work!"). It's an important distinction: Is time allotted by God for our moral uplift, or do we take it (or save it) for our economic and personal betterment?

There is another way in which Emerson's essays, Thoreau's *Walden,* Hawthorne's journals, or Melville's tales are affecting elegies for a passing America. The worldly seafaring tradition of New England had fed Thoreau's imagination in his months of self-exile. During his year at Walden Pond, Thoreau, an intellectual mariner, was never cut off from Western, even universal, culture. He read the Hindu devotional, Bhagavad Gita, he considered Greek and Latin classics, he imagined (if only to dismiss) voyages to every part of the wider world. The globe and all of history were (almost literally) at his back, and they were, to put it simply, his comforters, his friends. The true terror was the prospects opened up by the railroad, already launching its great westward journey into the darkness of unsettled territories, and in the process draining New England of its traditional leadership role. Portland, Maine, in the 1850s and sixties was larger than either of the emerging metropolises of Atlanta or Houston. For the first time in history, having the sea instead of empty spaces at one's back was an economic deterrent—or at least a less certain guarantor of future growth. The railroads were spreading and the oceans were their only barriers. Abundant, inexpensive, fertile, unpopulated land became more predictive of future success than a safe harbor, orderly villages, settled societies, noble history, books, and intellectual graces.

Reading the American transcendentalists, in fact, might lead

one to think that British Romanticism, chased out of its homeland by the Industrial Revolution, had found a refuge in New
England. Britain had embraced technological change so avidly
and successfully that nearly every miraculous invention of the
first half of the nineteenth century—chemical dyes, the railway,
the telegraph, the open-hearth oven—has a British provenance.
The decade of the 1850s in England commenced with the opening of the Great Exhibition of 1851, and the leadership provided
by that most progressive, most scientifically educated of all royal
patrons, Prince Albert. Profits from the Great Exhibition at the
Crystal Palace were plowed back into the building and mainte-
nance of technical colleges. This was Britain's answer to the
universal call to revolution: more industry, more science, more
research, broader technical education for the middle classes, science lectures to workingmen's clubs. The decade ended on the
publication of the most influential book of the nineteenth century, Charles Darwin's *On the Origin of Species,* which became
an instant best-seller. But, should I not have made my point,
Britain began that decade not just with the Great Exhibition, but
with the uniting and vitalizing introduction of standard time.

ANYONE LOOKING to the future of the two great English-
speaking cultures at that moment in history would have predicted
for England scientific, material, and economic dominance, under
a tone of unwavering rationality. For America, the same futurist
might have sketched a path of dreamy agrarianism and genial
mysticism with solipsistic tendencies. Given the apparent bent of
their national genius, it is striking that pragmatism, that most
American of philosophies, the test of validity through life experience, should have arisen in New England a mere forty years later.
Something revolutionary must have intervened to turn America
away from the New England model and in the direction of heavy
industry and technology, and the creation of new models of leadership, ambition and success.

One such "revolution" was the Gold Rush (1848–49), which focused attention on California and the fastest way to get there before the wealth ran out. Clearly, this left New Englanders at a disadvantage. One can hardly imagine a stronger contrast to the communal democratic traditions of New England than the golem that emerged from those raw, instant settlements in the Sierra foothills, or in the ports of San Francisco and Oakland. From them arose all the antisocial, anti-intellectual, vulgar, violent, and materialist impulses that America's founding elites and Tocqueville in his darkest moments had had good reason to fear.

During the next decade, New England took the moral lead in the antislavery debates, but the Civil War (1861–65) cut across the subculture of New England like an ax blade, severing the region, almost uniquely, from direct contact with the war or territorial participation in it. The war brought suffering and death to America on a scale never before or later seen, and along with it came authentic, one might even say existential, issues of sin and redemption in the real world, not in the gothic shadows. The war opened up the Midwest, unleashed industry, and elevated to leadership a new breed of impatient, self-educated, practical materialists, many of them the mechanical men of Thoreau's nightmare. The railroad men not only wore the railroad on their backs, they seemed to have ingested it as well. The generation of Harvard Unitarians was finished.

During those twenty years (1850–70) of the testing of the American identity, England was enjoying unchallenged hegemony over the world. "Sleeping giants" were everywhere in the 1860s—Russia, Germany, and the Japanese Empire notably, and there was always the uncertain fate of the chronic "sick man," the Ottoman Empire, which still controlled the restive Balkans, the Middle East, and much of North Africa—but the true sleeping giant was America, and it was only beginning to stir.

After 1869 in the United States, with the linking of the continent by rail and with the rapid populating of the formerly inaccessible West, the inner rhythm, that inherited sense of ordered time and space whose defining engines had been the horse and the sail metered by the sun, was ripped apart, never to return. Three thousand miles—a six months' journey around Cape Horn to the goldfields of California in 1848—could be covered in five days inside a single comfortable car. Settlements on the rail line, like Omaha or Denver, became major cities virtually over-night. Chicago quadrupled its population, all because of the rails. The integration of various networks, particularly the rail-way and telegraph grid (they were inseparable; railroads could not function without telegraphic signaling), demanded coher-ence and convertibility, and a simplification of the various time standards. Not incidentally, the first proposal for standard time, a way of simplifying the hundreds of bewildering railway sched-ules, came from a professor in Saratoga Springs, New York, Charles Dowd, and it is dated 1869.

Since North America extends five solar hours from New-foundland to the Pacific, and four more to the tip of the Aleutian chain, it offers the possibility of hundreds of viable time stan-dards. Time had been complicated enough when the majority of Americans were clustered along the eastern seaboard and had to negotiate time standards that were rarely greater than half an hour apart. Projecting the same complications across the entire continent was enough to induce temporal nausea. The territory was too vast, and the population growing too numerous—time had lost its meaning. Coordination of command is the reason armies run on military time, or, in our day, that airlines commu-nicate in a single world standard called "Zulu." Out on the fron-tier in the nineteenth century it didn't matter to cowboys what time it was on the cattle drive, or when they hit Dodge City for a drink; but if desperadoes planned to rob a mail train, or knock over a bank when the payrolls arrived a few towns away, a crude temporal calculation was necessary.

————

FLEMING'S ARRIVAL as an eighteen-year-old in the un-
promising village of Peterborough, Ontario, where tree stumps
still cluttered the main street, echoes the arrival of his hero, Ben
Franklin, in Philadelphia a century earlier. No position, no parents, just a strong back, a keen mind, and a willingness to work,
ready to do anything honorable that returned a decent wage.

As a result of Franklin's example, Fleming had probably not
wasted a minute without profound self-recrimination. From his
earliest diaries, even as a teenager, there emerges a self-portrait of
overall gravity, and an expectation of return on invested time.
Idle moments in his childhood and adolescence had been spent
in chess, sketching and hiking, and dreaming up inventions. On
a typical day in April 1843, at the age of sixteen, he sketched a
planned monument to Adam Smith. Then, in a scheme worthy
of Sir Walter Scott or Robert Louis Stevenson, he devised a
"fine exercise to make a plan of some old castle then fill up
the building and make it like the supposed original, such as Sea-
field, Peathead, Macduff's (with caves). One of the caves at the
Wemyss would make a fine place for drawing a band of robbers."
The afternoon was taken up learning Odell's Shorthand Alphabet, then studying the recipe for oil paper and another for case-
hardening. He sketched a church, designed a roller skate, then
copied an extract from *Poor Richard's Almanac*: "But thou dost
love life? Then do not squander time, for that's the stuff life is
made of. How much more than is necessary do we spend in
sleep, forgetting that the sleeping fox catches no poultry, and that
there will be sleeping enough in the grave. Sloth maketh all
things difficult, but industry all easy; and he that riseth late must
trot all day and shall scarcely overtake his business at night; while
laziness travels so slow that poverty soon overtakes him." Before
closing for the night, he described a machine for taking portraits
by the silhouette method, then designed a system of pumps for

propelling vessels at sea on the principle of Barker's Mill. (The "Barkers" were charged with removing bark from logs by directing high-powered jets of water.)

Physically robust, he led a full partying life; hangovers and morning-after flagellations are part of the youthful record, and were never really abandoned. In later years, he was forever retiring to the decks for a smoke, to his study for brandy and correspondence. His grandchildren remembered even in his old age the smell of cigars and brandy on his breath. Back in Kirkcaldy there was a lass, Maggie Barclay by name, with whom he might have settled, had he stayed. At least John Sang thought so. His letters rarely failed to mention her unchanged beauty and availability.

In 1843, at age sixteen, he sketched plans for an electrical-storage and light-distribution system. Two years later he wrote in his diary:

> I have been thinking for some time that the charcoal lights of the Magnetic battery might be brought to some practical use. I only require one experiment, but it would be an expensive one for me unless I could meet with a powerful battery, but I don't think there is one in Canada. It is to try if more than one light can be formed with one set of wires by breaking the connection and interposing charcoal points. If this is the case we have a good and cheap substitute for gas, would give a much better light and at least could be easily adapted to lighting streets or churches just by having a wire like telegraph ones with a charcoal appliance here and there.

Venture capital was an undeveloped resource in Canada and Scotland, and the prototype was never built.

And what about Sandford Fleming's Canada in 1848? It was a rough-hewn place. Montreal was the center of power, culture, and commerce, Quebec City the effective capital. Both conducted themselves as "English" cities. Toronto chafed under its comparative status with Montreal for the better part of a century. ("Hog Town" was Toronto's dismissive designation, uptight, moralistic, grim, and prissy, until the tide, a virtual tsunami, started turning in Toronto's favor in the 1960s and onward.) Fleming was then a young man of twenty-one, busily constructing a life, taking his town and harbor surveys, selling his maps, bringing his parents over and settling them on a farm near Lakeview. The outer world barely left an impression. On March 20, 1848, he noted in his diary: "News of the Revolution in France comes to Peterboro this morning." End of story.

It was not revolution—more, in fact, something closer to the temper of that time and place, a conservative counterrevolutionary riot—that greeted Fleming in Montreal in April 1849. He'd gone to the metropolis three months after his twenty-second birthday to sit for his surveyor's commission exam, but arrived on a night in which the Houses of Parliament were being torched. He and three young passersby rushed inside the lobby to save the portrait of Queen Victoria. ("Tonight," he reported, "I slept with the crown.") In 1850, at twenty-three, building upon two stimulating years as a member of a Toronto debating society ("Whether India or Africa Suffered Most from Europe"), he and a friend started the Canadian Institute. At the first meeting, they were the only attendees. Rather than abandon the dream, they elected officers: Frederick F. Passmore the president, Fleming the secretary-treasurer. A week later, he read a speech to scant attendance; a week after that, a paper on "The Formation and Preservation of Toronto Harbour"; and soon attendance started growing.

As a good Scots Presbyterian, Fleming was usually full of high-minded New Year's resolutions, transcribed in perhaps a shaky hand after a day of heavy "first footing" excess. In 1853, the twenty-five-year-old striver offered a glimpse of his busy life, and a view of the future:

Nothing can be recalled, what is passed not even a second ago, every action is as it were recorded on the minute of time for ever and ever. Do not regret the time I have spent (although I deeply regret other things) and the zeal shown in bringing into existence and into active operation the Canadian Institute because I believe it is calculated to do great good to my adopted country and to begin the New Year I have now resolved to provide it an endowment of £1000—when all that is mortal of me returns to the Mother dust, the interest of which to be annually expended in furthering the object of the Society. To effect this object I have already taken steps to assure my life for that, and may the over ruler of all things enable this humble creature while he lives to lack no opportunity in carrying this scheme out as cheerfully and as easily as it is now commenced.

The problem facing Fleming—and other Canadian visionaries wishing to import the technologies of Europe and the United States, as well as parliamentary-style, confederated government—was the lack of a forum. (It brings to mind a joke that Canadians hear all of their lives: Ah, blessed Canada! She could have had the technology of America, the government of Britain and the culture of France. What she got was American culture, French government, and British know-how.) Creating a forum for social and scientific ideas lay behind his establishment of the Canadian Institute. Elevating the political discourse, spreading scientific discovery, and making the scattered British colonies more representative and less dependent on Britain was a continual challenge. Too few people were spread over too great an area, segregated into too many competing jurisdictions, and further alienated by religion, language, and cultural tradition, for effective joint action to be anything other than fitful and self-defeating. Fleming was by no means alone as a progressive, but his efforts were often thwarted by the barriers of scale.

Nevertheless, he persevered. More than nearly anyone in the 1860s, he built the political support for the confederation of the separate British colonies, and succeeded. In the 1870s and early 1880s he orchestrated world support for standard time, and in 1884 he succeeded. Still later, and at great personal cost, he bravely rallied the informal commonwealth of overseas dominions against British communications monopolies, and succeeded in laying the worldwide, undersea cable.

A thousand miles east of Toronto, in 1851, the Railway Minister of the Nova Scotia colony, Joseph Howe, ventured a prophecy that is almost eerie in its accuracy. Fleming heard it as well, and filed it for later use. Fleming and Howe would not meet for another thirteen years, by which time Howe's obvious capacity for leadership would carry him to the top of Nova Scotia's parliamentary ranks.

Think you that we shall stop even at the Western bounds of Canada, or even at the shores of the Pacific? Vancouver Island with its vast coal treasures, lies beyond. The beautiful islands of the Pacific, and the growing commerce of the ocean, are beyond. Populous China and the rich East are beyond, and the sails of our children's children will reflect as familiarly the sunbeams of the South as they now brave the angry tempests of the North. The Maritime Provinces which I now address are but the Atlantic frontage of this boundless and prolific region—the wharves upon which its business will be transacted, and beside which its rich argosies are to lie. I believe that many in this room will live to hear the whistle of the steam engine in the passes of the Rocky Mountains, and to take the journey from Halifax to the Pacific in five or six days.

4

Time and Mr. Fleming

During the last fifty years, this new birth of time, this new Nature, be-
gotten by science upon fact, has pressed itself daily and hourly upon
our attention, and has worked miracles which have modified the whole
fashion of our lives.

—THOMAS H. HUXLEY, "The Progress of Science" (1887)

ABOARD THE *United Kingdom,* a steamship plying the North
Atlantic between Quebec City and Glasgow in May of 1863,
thirty-six-year-old Sandford Fleming, bearded, frock-coated as
always, paces the 235-foot deck, twenty circuits three times a day
for his daily three-mile constitutional, smoking his cigars, cheerfully conversing with captain, crew, or passengers. His long night
in an Irish country railstation lies thirteen years in the future.
The Canadian Pacific Railway is barely a fancy at this time.

His early achievements in surveying, land investment, map-
making, and his various civic ventures like the Canadian Institute
have all prospered according to his forecasts. He has engineered
one railroad, the thirty-mile Northern. Toronto is still his home.
Ottawa, the future capital of a vast dominion, and the city with
which Fleming is most identified, is little more than a canal land-
ing in 1863, named Bytown for its founder, Colonel By. Fleming

will build a mansion, "Winterholme," in the Adam style in Ottawa, where he will deed a park and a tropical arboretum. That, too, is in the distant future. In 1863, the vast dominion does not exist. Most of it—the Northwest, the area stretching from the Arctic to the American border, and from Manitoba to the Rockies—had been ceded by the British Crown to the Hudson's Bay Company in 1676, and the company still controls it. The maritime east is comprised of three jealously separate British colonies, Nova Scotia, New Brunswick, and Prince Edward Island, each controlled by its own elected leader, with separate laws and postage stamps—four, if the poor relation Newfoundland is included.

There was not yet a country nor a capital, but 1863 and (especially) 1864 would be crucial years for the maturing of Sandford Fleming and for the idea, if not quite yet the reality, of Canada. In 1864 Fleming would travel over twenty thousand miles for the cause of Canadian unity, some of it in comparative luxury, but much of it in hundred-mile-a-day horse-drawn sleds through the snows of Quebec and New Brunswick. Exactly half of his life had been spent in Canada. Married eight years, he was the father of four (on the way to seven), secure in his reputation as surveyor, an engineer, and a young man of enormous civic involvement, energy, and idealism. He had begun to take on duties that would define his next fifty years. In 1863 he had been given the opportunity to put the two halves of his life, Scotch and Canadian, back together. In the course of his voyage to Scotland, England, and Ireland, he became a public man, a visionary. This was his first trip back. Forty-three more would follow.

The voyage began on a familiar note, Fleming taking moral instruction from every aspect of nature and industry, projecting the heavy, elegiac mood of melancholic uplift. Passing under the Victoria Bridge in Montreal, on the way to Quebec City to meet the sailing of the *United Kingdom*, he noted in his travel journal, which he kept for the benefit (moral) and entertainment of his children:

How few in passing through now think of the men who built it, the men who planned, who laid these stones and drove their rivets, who think now of the anxious days Hodges spent when the ice was hourly expected to carry away the scaffolding under the center tubes; who dreams now of that army of skilled workmen who crossed the broad Atlantic to erect a national monument for Canadians, a monument that even if it does not pay six percent is far more useful than the pyramids and who asks how much percent they pay?

The engineering profession, always a high calling—and often a source of profound despair—for Fleming, is the link between science and society. The engineer calculates the cost of change, understands debentures and interest rates, the politically possible, the socially beneficial. He reads the future.

The *United Kingdom* steamed toward Glasgow (he noted) at nine miles an hour, burning eighteen tons of coal a day. Each morning's brisk circuit with the captain elicited new bits of information, speed and direction, weather and navigation problems. Collisions at sea were deemed improbable, icebergs, oddly enough, a more legitimate threat. He illustrated his journals, sketching whales, fellow passengers, the wrecked hull of a fishing boat floating off the Grand Banks, and smokestacks of distant steamers. The various personalities of Sandford Fleming were all taking shape: engineer, visionary, Canadian, Scot, patriot, networker, Empire Loyalist.

He was on a mission, carrying a petition to the Colonial Office in London. The (mostly) pioneering Scotch settlers in the Red River Colony, near today's Winnipeg, had chosen him—he was one of them, after all, as well as a reputable surveyor and railroad builder—to petition London for a railroad link from Upper Canada (Ontario) to relieve their isolation. The only way in or out of the Red River was by the ever-attentive American rail service from St. Paul, or river steamers "down to Canada" (the Red being one of North America's few north-flowing rivers) to the Canadian border. Thereafter, Canadian river steamers or ox-drawn wagons moved goods and emigrants to the colony. Buffalo herds, fording the rivers, could still force steamers to tie up for a day at a time.

The colonists had sound geographical reasons to feel isolated. The Canadian Shield, the oldest geologic structure on the planet —the granite nubs of ancient mountain ranges, pitted and pocked with lakes and bogs—is an effective barrier to any east-west communication. Beginning less than a hundred miles east of the colony, the land turns swampy and forested, leading to hundreds of miles of bare granite, mosquito-infested lakes, and unsounded tamarack sloughs. A few rough trails had been hacked out by fur-trappers and *coureurs du bois* over the years, following the higher and drier ground, but canoes were the primary form of transportation. The segmentation of Canada around Lake Superior was profound and seemingly inescapable, nothing like the minor annoyance posed by the chain of picturesque hills called the American Appalachians.

Geography and isolation, or fear of annexation, and skirmishes—often bloody—with the Métis (mixed-race French-Indian settlers of the Northwest who were militantly opposed to falling under Upper Canadian authority) were not the only problems. They were not even uppermost. Essentially, the colony could not be linked by rail to Upper Canada until the Northwest Territories, still part of the original Hudson's Bay land grant, could be legally transferred to an entity called Canada, and Canada did not yet exist. Lower Canada, the French-dominated Quebec, would hardly agree to the vast enlargement of English-speaking Upper Canada. In other words, the vulnerability, loneliness, and isolation of the Red River colonists—their essential "Who are we?" that defines the Canadian experience—mirrored the problems of British North America in general. Britain no longer controlled it, no longer even wanted it, and was anxious to turn over as much self-government and costs as the Canadians were able to assume. But Canadians felt the need for British protection on the American-dominated continent. In particular, Britain did not want to involve herself militarily or financially with the defense of Canadian sovereignty against the growing threat of American annexation.

In 1863 the Americans were locked in the most crucial year of the Civil War, and already eyeing the British colonies as hostile territory, ripe for annexation. Britain, given its need for foreign cotton, and chary perhaps of the North's rising industrial might, clearly favored the Confederate side. Many Northerners suspected that Confederates or their sympathizers might conduct raids on Union troops from Canadian havens. Such attacks from Canadian soil, Fleming and others feared, apart from violating neutrality, could be seized on as a provocation for a preemptive invasion, especially by an activist ideologue like the American secretary of state, William Seward. Fleming's English friend J. W. Wood wrote from London: "The North, whether right or wrong in the whole question of the war, have a good right to demand that our territory shall not be made a basis for such attacks upon themselves."

Seward, a leading advocate of "Manifest Destiny," held the notion that the United States was not merely a continental power but destined by its dynamism and the full exercise of its republican virtues to *be* the continent, as his 1867 purchase of Alaska from Russia would soon bear out. He did not disguise his designs on Canada and several of the Caribbean islands. Just a few years after the war, during the Ulysses Grant administration, noted Henry Adams in his *Education:* "[Adams] listened with incredulous stupor while Sumner [Senator Charles Sumner, Chairman of the Foreign Relations Committee of the United States Senate] unfolded his plan for concentrating and pressing every possible American claim against England, with a view of compelling the cession of Canada to the United States."

There is no evidence, incidentally, that any but a small minority of British colonists in North America shared the mother country's sympathies. Forty thousand Canadians volunteered for service on the Union side. Before the year was out, however, Fleming raised a company of seventy home guards in Toronto to repel a possible invasion. Thankfully, his skills and those of his neighbors were never tested.

ONE OF THE oddities of his voyage aboard the *United Kingdom* that he found amusing, and which he felt his children would especially enjoy, was the ever-earlier lunchtime occasioned by the ship's rapid progress northeastward toward Glasgow. He figured that lunch was being served half an hour earlier each day, by stomach time, due to their eastward travel. Always the instructor, or the Victorian father, he tried keeping his children close to him as best he could in "real" time by coordinating the North Atlantic time with that of Toronto:

In the evening the passengers as usual amused themselves, reading, playing whist and drafts. On deck about 11 o'clock enjoying a smoke with the captain who is most vigilant and is a little anxious about a fog coming in although there is no danger to be feared except a collision with another vessel— and the chances of this are exceedingly small seeing even in daytime and in clear weather we seldom see any other draft when off "the Banks." Left the Captain on deck with a good southerly wind and the ship scudding through the water at about 11 miles an hour, and thus went the eighth day aboard.

Wednesday 20 May. Up about seven o'clock ship's time and had an hour on deck before breakfast. The ship going as I left it last night with the same wind and from the same quarters. The fog has disappeared. The air is genial and moist, the wind strong but not cold, no ships were seen during the night and not a thing but the blue foaming sea within the scope of our vision.

Neither yesterday nor today did we get an observation of the sun, the sky being cloudy, but from dead reckoning the position of the ship today is lat. 49′25″ long. 39′32″ compared with that of Toronto about long. 79′20″. It will be seen that we are about 40 degrees nearer Greenwich and therefore 40/360ᵗʰ or about 1/9ᵗʰ the circumference of the Globe (in this latitude) away from Toronto and therefore our time on board should be 1/9ᵗʰ of 24 hours faster than Toronto time, hence where it is noon by my watch it is 2:40 o'clock by ship's time. Between noon yesterday and noon today we ran easterly about five and a half degrees longitude and as each degree of longitude is equal to 24 hours divided by 360, or four minutes of time, the ship clock has to be moved forward four minutes for each degree of longitude passed over and thus today five and a half times 4 equals 22 minutes. At this rate our lunch which is at noon will come about twenty minutes sooner every day. The distance between Quebec and Glasgow by the course of the ship is about 2900 miles and this afternoon we have passed over about half that distance. We have now been eight days out but we hope to gain a day on the remaining half and in that case may get into port a week from today.

He obviously knew how to compute the time, although his method was, as might be expected, a curious mixture of the natural and rational. He was computing the nautical day, "natural" time at sea, yet not making the "rational" leap into mean (averaged) time. Standardization simply had not yet become an urgent matter, except with regard to the floating lunch hour.

When he landed in Glasgow, he ran into a bootblack, "wee

Robert Gordon," whose pert, aggressive ways sold him on a shine. He also hired the boy to take him on a walking tour of Glasgow.

> I invited the little boot-black to breakfast with me which he gladly accepted. He sat opposite me and we had a long chat. He supports a widowed mother and earns from 8/- to 12/- per week. He had two boiled eggs, one cup of good coffee, two thick slices of bread and butter for the sum of tenpence or fivepence each. Robert then took me from Jamaica Street where we breakfasted along Buchanan and Argyle Streets to see the shops and I left him at the Exchange to pursue his vo-cation. I found Robert an intelligent boy and one who is bound to succeed in life.

No, there followed no promise of sponsorship to Canada, a job, or a college scholarship for wee Robert Gordon. There is simple ("manly," as Thoreau might have put it) recognition of a lad, much like himself, whose pride and genius for hard work, for responsibility, for individualism, recommended him to the future, another spiritual son of the patron saint of all self-made entrepreneurs, all young men of pluck, industry, native intelli-gence, and cultivated good luck, Benjamin Franklin.

FOR THE FIRST TIME in his adult life he was back in Britain, and for the very first time in London. It's hard to think of Fleming as a provincial, this man who would eventually hold the whole world in his gaze, linking it by time and by cable—but he was just that, and the effect on him must have been unsettling. London was the imperial capital, the center of the world. For twenty years, while England and the United States had been un-dergoing their very different political and industrial revolutions, Canada had been isolated, intellectually incubated, lacking dynamism, lacking a culture of its own. Fleming's avid interest in London, at the age of thirty-six, is more like a schoolboy's first encounter. He walked fifteen miles his first day, and kept up his journal to his children:

> It would be impossible for me to describe here the richness of the architecture of the buildings or all that is seen in passing through the streets of London; it is perfectly bewildering to attempt to notice everything and it would be endless writing to record all that comes before the eyes or the impressions found on the mind, everything is on a magnificent scale, distances, wealth, pomp, poverty and crime are all here developed to a greater extent than perhaps in any other portion of the world . . . we saw on all sides an ocean of buildings, disappearing in the smoky distance with scarcely anything for the eye to rest upon but chimney tops and church spires, to the east, the Town, and to the west [sic], St Paul's Cathedral.

And what must it have been like for a boy from the colonies, even a thirty-six-year-old boy, accustomed to the basics of Scotch-Canadian cookery, to be confronted by a High Victorian repast such as was served to him as a guest at the Civil Engineers Annual Dinner, on June 10, 1863? He thought enough of it to save the menu. From six-thirty till eleven-thirty, they ate and apparently ate some more. The menu reads like something from *Tom Jones,* and brings to mind old pictures of salmon catches, buffalo hunts, pigeon-snaring, and big-game safaris. This was London at its imperial height. One can imagine the mirrors and chandeliers, the hordes of waiters, the cigar smoke. They were addressed by Mr. Gladstone, chancellor of the exchequer, then by the Lord Mayor, then the Earl of Caithness, and, as Fleming put it, "a serene Highness of some description from the Continent."

FIRST SERVICE

Green Peas Soup   Ox Tail Soup   Mock Turtle Soup
Salmon   Whitings   Turbots
Broiled Salmon au Sauce Piquant
John Dory à la Hollandaise   Red Mullets en Papillote
Côtelettes de Saumon à l'Indienne
Stewed Eels   Trout   Soles à la Normandie
Whitebait

SECOND SERVICE

Entrées
Friandeau de Veau à l'Oiselle   Kari d'Homard au Riz
Côtelettes d'Agneau aux Épinards
Côtelettes Mouton aux Concômbres
Ris de Veuu aux Tomates
Poulet à la Marengo   Suprême de Volaille

Forequarters of Lamb   Saddles of Mutton
Roast Capons aux Champignons   Boiled Pullets à la Financière
Bacon and Beans
York Hams   Côte de Boeuf à la Jardinière   Ox Tongues
Roast Chickens   Veal Olive Pies   Pigeon Pies   Boiled Chickens
Asparagus   Cauliflowers   Salads   New Potatoes

THIRD SERVICE

Quails   Leverets   Guinea Fowls   Ducklings   Goslings
French Beans   Mushrooms   Green Peas
Prawns   Lobster Salad
Cabinet Puddings   St Clair Puddings
Gâteaux   Jellies   Creams   Meringues
Charlottes de Fraises   Richmond Maids of Honour
Pastry   Tarts
Omelettes aux Confitures   Orange Fritters   Nesselrode Puddings
Wines and Liqueurs
Sherry   Madeira   Hock   Champagne   Sparkling Hock & Moselle
Old Port   Château Lafitte
Curaçao   Maraschino   Eau-de-vie   Usquebaugh
Dessert-Coffee

His return voyage, on the *Great Eastern,* brought him to New York (from Liverpool) on July 1, 1863, a date that would mark the birth of Canadian Confederation in four more years. On July 4, the American passengers staged an impromptu celebration of their nation's birthday. He was teased into carrying an American flag at the head of a deck parade, to which he agreed on the condition that an American would similarly honor the Union Jack. That accommodating American, the Irish-born William Dawson, later became the mayor of St. Paul, Minnesota, and a lifelong friend. Fleming was always sociable, helpful, and approachable; he was particularly adept at what these days is called "networking."

The return crossing had taken only eight days from Liverpool to the first North American landfall, Cape Race, Newfoundland—one-sixth the time of his original sailing in 1845. But as the ship headed south toward New York, heavy fog set in, as it often did off the Grand Banks and along the New England coastline, requiring careful navigation and adding an additional ten days to the journey before the arrival in New York Harbor.

Fleming was elected ship's historian for the voyage, a custom of the time to celebrate the passage in some witty and charming manner for publication in the local papers. He was still writing when the ship entered Long Island Sound; still writing when the pilot-crew came aboard spreading news of a Union victory at Vicksburg, and, as he put it, "that General Lee was having the worst of it in his invasion of the states north of the Potomac." The passengers gathered on deck to congratulate the captain and to praise the ship. Mr Fleming, ever the engineer, ascertained that the ship had burned three hundred tons of coal and now stood three feet higher in the water than it had in Liverpool. Then he handed over his article, which was published in the *New York Herald* the next day under the headline NEW CANA-DIAN PASSAGE PROPOSED in the Marine Affairs column. In its way, it is as remarkable a piece of prophecy as Joseph Howe's had been in 1851, predicting a trans-Canadian railroad with nothing but the Atlantic Ocean at his back.

What Fleming proposed was closer in spirit to his two major future projects, standard time and worldwide cable, than to any of his accomplishments to date:

> Twenty years ago from five to seven weeks was considered a fair passage across the Atlantic, and although much has already been done through the instrumentality of science and iron and steam to destroy the terrors of an ocean voyage, it requires no effort to perceive that much more must be accomplished before the line of passenger traffic between Europe and America is perfected.
>
> We must have more *Great Easterns* and the time at sea must be reduced to the minimum number of days. Half the time we have spent on board this magnificent ship has been occupied in coasting and I cannot understand that the owners of the vessel can be any better disposed to keep us at sea than we are to remain on dry land. I do not here speak for myself but for the generality of travellers as I rather like a sea voyage when time admits, but it seems very clear that the ocean voyager will ultimately be confined to the shortest duration between land and land.
>
> The great bugbear has always been the length of the sea voyage and sea sickness which has hitherto accompanied it. Now length of the voyage would be diminished one half if a proper land communication existed between the eastern coast of Newfoundland and the railways of America. Seasickness barely finds a footing aboard the *Great Eastern.* I believe the doctor of the ship could report that there has been less sickness of any kind amongst the 1500 souls on board than generally exists in any town of the same population.
>
> Now distance between Ireland and Newfoundland is less than 1700 miles which at the rate of sixteen miles an hour

would require four and a half days to run it. The *Great East-ern* runs it without any effort in five and three-quarter days and considering the improvement which can be made in speed, I feel sure that allowance of five days for the ocean voyage would be ample. With regard to connecting St John's or some equally good harbour on the Atlantic coast of New-foundland with the railway system of the interior a glance at the chart will show that the most direct course is to traverse Newfoundland by a railway 240 miles in length to the Gulph of St Lawrence, thence by steam ferry (about three times the length of the ferry between Holyhead and Dublin) to Gaspé, thence by an extension of the Grand Trunk Railway to the interior of the United States to New York and to Canada. To establish this route the construction of some four hundred miles of railway would be necessary and beside a sufficient number of ocean steamers like the *Great Eastern,* powerful steam ferry boats to cross the Gulph of St Lawrence at all seasons would be required.

Let us glance briefly at what would be accomplished by establishing such a line for traffic on the scale indicated. The journey from New York to London could be done in seven and a half days, and from Chicago to London in eight days while the ocean passage would be reduced to five days which performed by steamers like this one would throw every other line at least of passenger traffic entirely in the shade. When such a passenger route is established the ease, speed and comfort with which the voyage could be accomplished would have the effect not only of concentrating traffic to this the shortest passage between the two continents but also of greatly increasing the number of travellers. It would in fact become the great highway between the old and new worlds and it is not at all improbable that the speed and comfort of the voyage would increase the traffic to such an extent that in a very few years a daily line of *Great Easterns* would be called into registration and thus the ocean would virtually be bridged by a system of steam propelled floating hotels.

Mr. Fleming had entered a new phase. He had delivered the Red River petition, and it had been respectfully received, but with the predictable recommendation that the Canadians should build their own railroad with London's blessing and approval. He had been accepted by his British and American peers, and he had been graced with a vision of Canada's future that could indeed make her a player on the North American scene. Of course, there was not yet a Canada. The contrast among the United States at war, Britain in its imperial full flower, and ragtag Upper and Lower Canada had never been more striking.

IN THE FALL and winter of 1863–64 he held a commission to survey the lands to the east and south of Quebec City, the colonies of New Brunswick and Nova Scotia, a feasibility study, we might say, for the building of the long-delayed Intercolonial Railroad from Quebec City to Halifax, should all go well. Such a railroad would link the ocean ports of Halifax and Saint John, New Brunswick, with the river ports of Quebec and Montreal, and the Grand Trunk Railway inland to Toronto and beyond.

Border disputes between the United States and the British colonies had rendered earlier British maps useless. Most of the territory originally surveyed for British development, including all of northern Maine, was now under American control. Fleming believed that the United States had never intended to claim, nor had expected, to gain northern Maine, and even if it had, the maps that had supported its claims were in error. Had England stood her ground, presented sound surveying evidence, northern Maine would have remained in British hands and the passage between Quebec City and Saint John would have been comparatively direct. Because of London's malfeasance, as he saw it, he was forced to loop a line north of Maine, adding hundreds of miles of track, and millions of dollars to the costs.

Fleming rarely took Americans to task for exploiting London's indolence wherever they could. It was the Colonial Office in London that had weighed the negligible costs to England of losing Maine against the possibly open-ended expense of defending it. Over the years, he became a strong supporter of Empire loyalty based on common history, culture, and instruments of government, but an even stronger advocate for the worldwide network of linked British states that came to be called the British Commonwealth.

Ever the multitasker, as we'd say today, Fleming was not merely surveying the hills and forests of New Brunswick for a rail line. He combined those duties with the political cause of colonial unification, the creation of a greater Canada to combine Upper and Lower Canada with the maritime colonies, along with the annexation of the vast holdings of the Hudson's Bay Company in the heart of the continent. With the United States remaining an active threat (as well as a temptation), he realized, along with others, that to save the many scattered parts of British North America it was first necessary to create a single entity, no matter how repugnant it might be to the various colonial premiers who would stand to lose their positions.

Fleming's professional mission as surveyor was combined with a political mission to the Charlottetown Conference, a meeting to discuss the issue of confederation, held in Charlottetown, Prince Edward Island, on September 1, 1864. On his return to Quebec, he brought along J. W. Wood, the conference's recording secretary, a young Englishman whose surviving file of letters is a moving record of an almost mystical period in Canadian nation-building. Mr. Wood's summer, which he spent in the forests, in wagons and on horseback, in canoes and on foot, in isolated villages and the gracious homes of its leading citizens in the company of Sandford Fleming, remained with him for the rest of his very long life.

On one level, it is a record of Fleming's effective networking, securing a political base for unification among the colonies' leading citizens. On another level, what comes through even forty years later in Wood's recollections is that those weeks in 1864 were a blessed summer in which he (Wood) leveraged his way, however modestly, into a significant chapter in British (never, to him, Canadian) history. He was present at the inception, if not quite the birth, of a nation. It's hard not to imagine Wood and Fleming as figures in a vast landscape painting, two young men on horseback following a muddy trail through a dark forest, a salmon stream to one side, purple hills in the distance, smoke rising from the chimney of a nearby cottage.

History does not provide official records of the Charlottetown deliberations. It comes as a supririse, then, to learn—through Wood's letters—just how fanciful, or perhaps merely enthusiastic, some of the ideas were. The most persuasive orator of the confederation cause, Fleming's close friend, D'Arcy McGee, proposed the unification of the various colonies of British North America under a prince of the English royal house and a daughter (if there be one) of the royal family of France. Fleming, according to Woods, endorsed the idea whole-heartedly:

[The proposal] would have the happiest influence upon the destinies of the North American Provinces. It would be received by the French portion of the people as a high compliment to themselves. It would bury, by amalgamation with their present feelings of allegiance to England, any still lingering memories of the land from which they have sprung. . . . At least it would appeal in a powerful manner to those sentiments which are so accessible to the French temperament and would increase immensely that feeling of common interest and common country which is so desirable to foster and develop.

Although no degree of mutual ignorance between the two founding peoples of Canada should be surprising, the idea that French-Canadians, by then a hundred years removed from French authority, remnants of a pre-Revolutionary French culture, Catholic heretics (Jansenists) raised on a dogma that characterized post-Revolutionary France as the devil's own breeding ground, would tolerate a pretender from the French royal house exists only on the furthest shore of probability.

Wood soon returned to England, and "the current of his destinies," as he called it, took him to India for the next forty years, building the Bombay, Baroda & Western. In the two years leading up to confederation, he had an opportunity to reflect on the impossibility, even the futility, of the Canadian project. Every now and then he expressed seditious thoughts, begging his old friend not to hold it against him. In 1866, from Bombay, Wood wrote:

Although I usually do not feel sure that it is of very great importance either to England or the world at large whether the great country now forming our territories in North America is thoroughly opened up in the British interest, or not, so long as it is opened up for the benefit of mankind, I, looking on as an outsider, (and I fear you will say an unsympathising one, though it is not so) can imagine a much worse fate befalling our North American Provinces than absorption into the United States, and I am inclined to think that this will come sooner or later. The U.S. will in time want the whole country, at least they will, after a while, become full-blooded again and agitate this and other questions, and though of course, while bound in honor to do so, England would fight the battle with the colonies in case of war, I cannot, I confess, see that such a risk is worth running either for the colonies or for England. . . . And, looking to the development of the unopened country, it seems to me comparatively a matter of indifference in the interest of humanity whether it is done by people called Englishmen or by people called American.

Wood faithfully reflects the English, not Canadian, view. The least confrontation, the least complication, the better. Canada was Britain's to give away, if necessary, not to nurture into independence.

If Canadians did not destroy their own prospects of confederation by petty jealousies or dynastic fantasies, the United States could be counted on to erect every barrier legally available, short of invasion, to derail it. But American pressures, economic and diplomatic, in many ways proved counterproductive to the aim, considered an inevitability, of annexation. It forced confederation and hastened the building of the Canadian Pacific Railway.

When Secretary Seward purchased Alaska from Russia in 1867, it was in great part to discourage Canadian unification and to limit the extension of a British coastline on the Pacific. Alaska, a much-ridiculed purchase ("Seward's Folly," "Seward's Icebox"), was not an end in itself but a strategic building block in the eventual Americanization of the entire North American landmass. (It might be said that the only thing that inhibited overt American action was an unchallenged belief in its inevitability, as shown in a *Chicago Tribune* editorial as late as September 5, 1884: "It is not necessary to discuss it in this country or to seek to force or even hurry annexation. All the elements—commercial, financial, social and political—are gravitating Canada towards the American Union, to which she naturally and geographically belongs.") Seward's powerful colleague, as chairman of the Senate Foreign Relations Committee, Senator Charles Sumner, termed the Alaskan acquisition "a visible step in the occupation of the whole North American continent." At the time of his exit from government, in 1869, Seward was negotiating with Denmark for the purchase of the Virgin Islands (which were finally bought in 1917, at some coercion) and Greenland.

It may seem bizarre today, but relations between Canada and the United States suffered through decades of rhetorical abuse, moments when Canada's historic fear of absorption matched America's frustration with the very idea of an alien, border-sharing presence. Throughout the 1870s and eighties, so long as Manifest Destiny remained and even grew as a living cause, faithful newspapers like the *Chicago Tribune* could be counted on to keep the temperature just short of boiling, should Canada persist in ignoring the inevitable:

> If ever there should be any serious misunderstanding between the United States and Great Britain, it would take but little time before an irresistible American army would cross the frontier, and annexation would be brought about so quickly that it would make the head of the Marquis swim [i.e., the Marquess of Lorne, the governor-general].

And, later:

> We are willing to put up with many things the Canadians do which we would not tolerate if the British were the immediate agents. The Canadians now divide with the United States the Great Lakes and the St. Lawrence River. We permit them to line our frontier for 3,000 miles with their custom-houses where American goods and American people are subjected to an Inquisitorial process; we permit them to disrupt our frontier; we submit to very many annoyances, with patience, because of the feeling that they are weak, and that it would be ungenerous in us to get angry with them. But let "the majestic" form of the United Empire intrude itself between the two

countries, and the United States may decide that the Nation should go in and twist the British Lion's tail.

Fleming's managerial history exposes another curious trait. Not just as an engineer, but in all aspects of his life until the standard-time movement, he was a curiously reticent, or ambivalent, leader. He allowed others to take the credit; perhaps we can say he was an engineer, that tragic profession, to his bones.

In 1872, Stanford Fleming was appointed Chief Engineer of the Canadian Pacific Railroad, the major financial undertaking of nineteenth-century Canadian history. He delegated authority to crew chiefs who later attacked him. When he cofounded the Canadian Institute in 1850, he chose the role of secretary-treasurer rather than the presidency. On his first surveying tour of the West for the Canadian Pacific Railway in 1872, he installed his closest friend, Reverend George Grant, as expedition historian, the result being Grant's classic account, *Ocean to Ocean*. As he gathered support for confederation, J. W. Wood wrote the letters, D'Arcy McGee made the speeches. At the Prime Meridian Conference—*his* conference, in many ways—he was "attached," like a satellite, to the British delegation as the representative of Canada, a country that did not exist. Even in his final years, rather than write an autobiography, he practically dictated a biography, *Empire-Builder*, to his friend Lawrence Burpee.

He often expressed himself on the engineer's calling as a kind of secular religion. Its codes were no less rigorous than the Hippocratic oath. In his view, engineering had a noble, almost tragic nature. In 1863 he'd extolled the profession in terms that recall the Book of Isaiah (every valley shall be exalted). "It is one of the misfortunes of the profession to which I am proud to belong that our business is to make and not to enjoy; we no sooner make a rough place smooth than we must move to another and fresh field, leaving others to enjoy what we have accomplished." By 1876 his views had only deepened:

Engineers, as you all know, are not as a rule gifted with many words. Men so gifted generally aim at achieving renown in some other sphere. . . . Silent men, such as we are, can have no such ambition; they cannot hope for profit or place in law, they cannot look for fame in the press or the pulpit, and, above all things, they must keep clear of politics. Engineers must plod on in a distinct sphere of their own, dealing less with words and more with deeds, less with men than with matter; nature in her wild state presents difficulties for them to overcome. It is the business of their life to do battle against these difficulties and make smooth the path on which others are to tread. *It is their privilege to stand between these two great forces, capital and labour, and by acting justly at all times between the employer and the employed, they may hope to command the respect of those above them equally with those under them.* [Italics added.]

Those last words proved as much prophetic as elegiac. Almost as though he had seen into his own future, he had delivered a judgment against the state of affairs in Canada, and his role on the Canadian Pacific Railway. He could no longer interpose himself between the demands of politics and capital, and his oath as an engineer.

Engineers are not paid to play Hamlet—if they must play the prince, better to take direction from Machiavelli. Parliament did not interest itself in engineering complications. The unsurveyed muskeg swamps between Lake Superior and the Manitoba border were found to be inland seas of gelatinous peat, capable of swallowing tons of sand and gravel and whole trains without ever reaching bottom. In one area, seven layers of rail lay buried, one upon the other. Three hundred miles of Ontario muskeg could absorb the personnel and resources—nine thousand men worked on the Lake Superior section alone—of two thousand miles of prairie track. There was no existing text, no standard of engineering to which anyone could appeal. Blasting, filling, and bridging through the Canadian Shield cost upwards of $700,000 a mile. The economical solution, sharing lines between Chicago and northern Minnesota, was of course politically unworkable.

The closing act in the engineering career of Sandford Fleming was his inability to decide on a proper route through the Rockies, the northern Yellowhead Pass (west of today's Edmonton), or the southern Kicking Horse. Unable to resolve it, he wasted years in surveys and second and third opinions, and a second personal survey. His indecision kept construction and surveying parties paid and provisioned, but often idle through the harshest winters on the continent. In that robust, high-wage, labor-scarce railroad-building era, it was difficult to keep crews together, especially when the chief himself appeared to be dithering in Ottawa and London. Some members defected, some complained to Fleming or to parliament. One built the street railroads of Oslo. Others left for India and the West Indies. Many more signed on with American railroads.

5

The Decade of Time, 1875-85

BY THE MIDDLE of the 1870s, the assertion of human reason over the processes of nature was yielding discoveries and inventions in all the arts and sciences that lent that famous Victorian confidence to the notion that man was no longer the passive inheritor of an ordained "natural" universe. All of nature was his to discover and mold. The ability to communicate instantaneously by voice, to light the dark, the luxurious trans-Atlantic steamers, the transcontinental railroads, a new personal printing press called the typewriter, bound the world in exciting and, for some, alarming new ways. But the outworn shell of time, those heavy boots inherited from tradition, from nature, were impeding progress. Societies were moving faster than their ability to measure.

Before railroads began serving every "civilized" part of the globe (as the Victorians were fond of calling it), the sun had set the temporal rhythm. Two cities set one hundred miles apart maintained an eight-minute temporal separation. But a train could cover a hundred miles in less than two hours—so which town's "time" was official? Which standard should be published? The train itself might have originated in a city five hundred miles distant. Who, therefore, "owned" the time—the towns along the route, the passengers, or the railroad company?

The burden was entirely on the passenger. Railroad compa-

nies owned the time. Upon entering larger stations where a transfer between lines might be necessary, American passengers would
study clocks set along a wall behind the ticket counter, each announcing the time standards of competing "roads." The clocks
would not read: "New York," "Chicago," "New Orleans," and
"Cincinnati," but rather, "Erie & Lackawanna," "New York Central," and "Baltimore & Ohio." Each separate time reflected the
standard at the company's headquarters. The Pennsylvania Railroad maintained Philadelphia time along its entire route, while
New York Central kept the "Vanderbilt time" of Grand Central
Station. If a passenger wondered when he might arrive at his final
destination, he had to know the time standard of the railroad that
was taking him there, and make the proper conversion to the
local time at his boarding, and that of his eventual descent.

For example, if you were a Philadelphia businessman in the
1870s with an appointment to keep in Buffalo, transferring in
Pittsburg (as it was then spelled), you would of course have to
know the departure time in Philadelphia local time (just as you
would today)—*unless* the train had originated in Washington or
New York, in which case it might depart according to the local
time of those stations, a few minutes earlier or later than your
local Philadelphia time. It was your responsibility to know the
difference. Thereafter, you entered a twilight zone of competing
times.

Pittsburg, five degrees of longitude west of Philadelphia, adhered to the sundial precision of its solar noon, which came
twenty minutes after Philadelphia's. The train you'd catch in
Pittsburg to take you up to Buffalo originated, however, in Columbus, Ohio, three degrees west of Pittsburg. That translated
to twelve minutes earlier than local Pittsburg time, or thirty-two
minutes earlier than the time on your still-uncorrected Philadelphia watch. A train arriving in Pittsburg from Philadelphia at five
o'clock Philadelphia time would find that it was only 4:40 in
Pittsburg (which was irrelevant, unless you were leaving the sta-

tion and staying in the city), but that the Columbus train would be arriving twelve minutes before that, at 4:28 Columbus time. And when you finally arrived in Buffalo (assuming you didn't miss your train), you'd be confronted with Buffalo's three official times, based on the three railroads that served the city—a philosophical conundrum that had, in fact, spurred Professor Charles Dowd's first serious attempt, in 1869, at temporal rationalization. For Dowd, the temporal conflicts created absurdity. Passengers were made hyperconscious of time, of each passing minute, in ways we cannot imagine today. For many passengers, it created anxieties bordering on agony.

No wonder Oscar Wilde, the serenely contemptuous child of British standard time, noted that the chief occupation in an American's life was "catching trains." No wonder Sandford Fleming cautioned against even the mention of "local time."

All of those times, in Pittsburg, Buffalo, and Philadelphia, were occurring at the same cosmic instant. But whose "now" are we talking about? It depended on what "now" meant. "Now" was composed of three, six, fifty, an infinity of separate times, all of them official, all of them accurate. Today, all three times are in the Eastern time zone; one time fits all. For the gentleman of the 1870s, however, right up to the moment of North American railroad standardization, in 1883 (a year before the world conference), all three times were legitimately (one might even say, morally) distinct. It was up to the traveler—he who would pierce the twelve-mile-wide bubble of local time—not the railroad, to make the adjustment.

The helpless passenger didn't yet realize it, of course, but his frustrations had already ignited furious debates behind the scenes. Railroad men, astronomers, grand theorists, diplomats, all quickened to notions of time and new ways of reckoning it. It was a reformer's cause, an opportunity to sweep away the residual pieties of a "natural" mind-set. Standard time, given the nature of its opposition from religious thinkers, agrarian traditionalists, and the contented, nontraveling public, became a popular symbol of progress and rationality.

Passengers were demanding something simpler. Proposals were raining down on the engineering profession, the railroad industry, the post and telegraph services. The American Metrological Society (measure reformers), the American Society of Civil Engineers, the American Railroad Association, all maintained "time conventions" to monitor their members' suggestions for purposes of advocating positions, placating the public, and thwarting political intervention. Some of the proposals they studied and even debated were ingenious. Some, in fact, like that of the very persistent Professor Dowd, were prescient. But the railroad industry, rather like today's Internet, was terrified of government intervention—lest public frustration boil over and attract political involvement—and loath to interfere with its own profitability and entrepreneurial independence.

Within the Decade of Time, the contradictions between new technology and old time-reckoning passed from inconvenience and inefficiency to urgency and, finally, to danger.

Ships of different nationalities could not communicate their positions at sea, due to competing prime meridians. Railroad accidents were daily events, an inevitability considering that trains on the same track might be employing different times. Meanwhile, the technology continued to evolve; the velocity of the culture continued to increase.

In that decade, just to name the most obvious examples, the telephone, the electric light, the typewriter, motion pictures and stop-action photography were invented. Even military misadventure could be turned to creative use. In 1871, upstart Prussia defeated France, throwing a proud culture into despair, but forcing a thorough self-examination and restructuring. From that fortunate defeat arose the determination to revolutionize institutions and to create new centralized and "rational" authority to cast off the dead weight of "natural" thought. France built a modern industrial state and unleashed energy that made Paris synonymous with art, culture, experimentation, and revolution. Baron Haussmann redesigned the medieval metropolis, replacing stagnant quarters with broad, railroad-style boulevards.

In the United States, because of their superior bogie-design, luxurious Pullman cars were carrying passengers from the Atlantic to the Pacific, offering greater comfort on the move than most Americans enjoyed in their homes. Sumptuous steamliners plied the Atlantic in little more than a week, which registered on veterans of the sailing days as instantaneous. Starting in Belgium and spreading quickly to Germany and France, Georges Nagelmackers's *wagon-lit* (sleeping-car) service, importing the bogie, was by 1883 serving chilled champagne and hors d'oeuvres of oysters and caviar followed by a full five-star Paris restaurant meal, with the same formal dress code, on the fabled Orient Express all the way from Paris to Istanbul.

By the end of the decade, then, ordinary man had become superman, speaking over distances, banishing the dark, chilling his wine and beer, and speeding through immemorial landscapes— the buffalo-dotted prairies, southeastern Europe—without regard to local time or outside conditions. Our modern world, for better and worse, was taking shape. Ancient empires were collapsing, particularly in the lawless Balkans. Many of the smaller towns along the Orient Express lines had to be bypassed, not that the passengers noticed, on account of firebombed stations or the murder of crossing guards. The imperial land-grab in Africa and Asia entered its sociopathic phase in the Congo, Tanganyika, and Tonkin, and el-Mahdi was chasing the British and poor Gordon Pasha out of the Sudan, destroying forever the early Victorian pretense of a "civilizing mission."

New fields of inquiry turned the techniques of stop-action photography of Eadweard Muybridge, who developed cameras with shutter speeds of $\frac{1}{200}$ and $\frac{1}{500}$ of a second, on human behavior itself. Sociology and psychology fragmented time into investigative frames, showing, through microanalysis, the irrational to be familiar, and the "normal" to be nothing less than bizarre. Individuals learned they were strangers to their own motivations; societies were seen as structured around prejudice, superstition, and irrationality. In American factories, Frederick W. Taylor introduced "scientific management," using the stopwatch instead of a stop-action camera to reduce the "natural" habits of laboring men and women to microanalyzable segments, with an aim toward improving productivity by replacing natural routine with rational efficiency. In painting, the impressionists broke with the careful perspective and shadowing of the Salon, the calculated posing and anecdotal portraiture, favoring instead bright shards of pure, unmodulated color, the painterly equivalent of stop-action. Impressionism is as much about time as it is about light. It's all about time.

Writers came later to change, as befits the reflective and reportorial nature of their art, but once they felt themselves in control of time, free to experiment with sequencing, able to shatter "natural" consecutivity, their works grew closer to the stop-and-go flow of consciousness itself. Temporal distortion became the surest way of communicating disturbance, urgency. Readers were stimulated into active involvement. Unbalancing the reader was not merely an instinctive political act, but a proper aesthetic tool for keeping the naked consciousness in focus, free of all that fussy Victorian decor. Conventional plotting, in fact, was regarded as mimicking the path of unconscious repression.

All these innovations and inventions derive from, or helped to create, an altered relationship to time that we call modernism. It's easy to trace the effects of time, a little harder to find the moment when it started. The grand events enumerated above all had their modest origins. Ross Winans's invention of the bogie; George Pullman's designing—and nearly over-designing for the track bed and existing gauges—the heavy funeral cortege for Abraham Lincoln; the idea for the *wagon-lit* coming to Georges

Nagelmackers when he was sent from Belgium to the United States to forget a failed love affair—only to fall in love with Pullman cars instead; van Gogh's seeing an exhibit of Japanese woodcuts in Antwerp and falling in love with their pure color and spatial foreshortening. Any number of artists saw Eadweard Muybridge's panel of stop-action photos of a galloping horse (and a genre of horse-racing paintings died on the spot). Other painters took from physiologist Étienne-Jules Marey's moving pictures the theory of the persistence of vision. There are obviously dozens of such moments, and cases can be made for each of them as the *ur*-moment in the birth of a new consciousness.

Central to all of them, however, is the need for a catalyst, an interactive but noncontributing agency, and that, I feel, is the set of preconditions that I've been calling the standardization of time: adjusting time to new velocities, distributing it equally, replacing nature with reason, religion with humanism. The logical places to look for it are London, Berlin, Vienna, and, of course, Paris.

Older and more settled, now with a country and many projects behind him, Sandford Fleming enters the scene, in one of the least prepossessing places in the world.

AT 5:10 P.M. on a bright, July afternoon in 1876, in the country station of Bandoran, situated on the main Irish rail line between Londonderry and Sligo, a balding figure with a salt-and-pepper mattress-stuffer of a beard and wearing a gentleman's formal frock coat, alighted with his international traveler's baggage from a horse-drawn taxi twenty-five minutes before the scheduled 5:35 P.M. arrival of the Londonderry train. Clearly, an important man, a distinguished visitor. Perhaps he read a book or a paper as he waited for the train; he was not one to waste a moment. But the station remained suggestively uncrowded, most unusual for a market town on the main line, as the arrival time approached. At 5:35 P.M. nothing came. He checked his *Irish Railroad Travellers' Guide* again, for he was in all matters metic-

ulous. There was no mistake. Perhaps he then inquired of a sta-
tionmaster, or scrutinized the departure board. It would read:
Londonderry 5:35 A.M. Sandford Fleming, chief engineer of the
Canadian Pacific Railway (CPR), would be a prisoner for the
night in Bandoran station, and in the morning miss his ongoing
connections to the ferry and to England. And in those hours a
plan slowly took form.

In 1876 Fleming was forty-nine years old, "well near the meridian," he'd noted on his birthday earlier that year, and hailing now
from that raw and aspiring capital of his not-quite nation, Ottawa.
His involvement with standard time, which was to begin that
night in a misprint, will fill and even define the "decade of time."
Other leaders in the standard-time movement, whom he will soon
meet and with whom he will correspond—and eventually lead
—were academic or naval astronomers, educators, or railroad
managers in charge of maintaining schedules. Fleming was more
practical than academic, and more theoretically inclined than
nearly all route managers. He was the government-appointed engineer-in-chief of Canada's two major railroad-building projects,
the Quebec-to-Halifax "Intercolonial" and the "national dream,"
the Toronto-to–British Columbia Canadian Pacific Railway. The
CPR was Canada's major financial undertaking, upon which the
survival of the nation was at stake.

Until Fleming's first paper, which followed the Bandoran misadventure by only four months, most proposals for time reform
had come from deep inside the American railroad industry and
had applied themselves exclusively to the reform of North American railroad schedules. No one knew trains better than Sandford
Fleming, but his proposals, from the beginning, treated the
needs of the railroads as incidental to the overall regulation of
time itself. Not only that, he paid no special attention to North
America. He was a theorist of world time. Before that day in Ireland, time had meant no more to Fleming than it had to most
busy Victorians. If he thought about the complications of solar

time, or of ways to repair them prior to his enforced stay in an Irish station, there is no evidence in his letters and journals. He'd simply soldiered on like most Victorian travelers, making rough adjustments at sea on his annual crossings to Britain, or on his surveying missions into the Canadian bush.

Missing that train, though it cost him sixteen hours and, as he put it, "monumental vexation," turned out to be the luckiest misfortune in his very lucky life. Had it happened a year or two earlier, it might have ended with an angry letter, or a snappish aside in his daybook. But in June 1876, while on a year's "medical sabbatical" from the CPR, he was a gentleman of unaccustomed leisure. And so, as he thought the problem over, he concluded that the error was more than a microscopic misprint. It was the opening into an industrial underworld of fallacy and inefficiency. Correcting it was not an editorial step; it was an abstract engineering project. The waste of his sixteen hours was a small instance of the world's daily loss from the retention of an outmoded system of time notation. Simple misprints were unavoidable and inevitable, of course—but in a larger sense, this particular error was *not* correctable, at least not under present conditions. The whole system of numbering was wrong; it alone accounted for the bedeviling misprint of the Latin abbreviation. Why should modern societies adhere to ante meridiem and post meridiem, why double-count the hours, one to twelve, twice a day? Are we so stupid that we cannot compute above the number twelve?

I N 1 8 7 6 Fleming himself was in a precarious position that perhaps even he did not fully comprehend, and certainly did not admit. He had begged from Parliament a few months earlier, and been granted, a year's medical leave. He'd cited a "near fatal" accident that had put him on crutches for several months. The accident from which he was recovering has never been identified, although a serious injury sustained on a surveying mission five years earlier had led to liver and bowel damage, opiate treatment, and two weeks in bed. Neither injury is mentioned in *Empire-Builder.*

His Ben Franklin–inspired worldview had never allowed for much in the way of doubt, hesitation, or morbid introspection. At forty-nine, the ever-upward, ever-busier public-service career of the most distinguished Canadian of his age was three years from ending, and his "sabbatical" would stretch onward for forty more years. Fleming's journals are useful sources for documenting travel, income, and expenses, the births of children and deaths of friends and family, but they very rarely indulge in self-analysis. The injury for which he requested recovery time does not appear to be physical in nature, but rather, a nervous crisis, a loss of confidence, perhaps the most wounding injury he could have suffered.

He was losing control of the CPR mission. It was a greater undertaking than any one man could master under the conditions imposed by Parliament, especially upon an engineer still involved with the earlier and smaller Intercolonial. He was a civil servant, answerable to elected officials, a chief in name only. Funding came from Parliament. His salary, though generous for the times, was that of a government official—and this in the era of the railroad barons, the Cornelius Vanderbilts and James J. Hills. Canada itself was only nine years old in 1876; the lines of parliamentary authority were still being drawn, and partisan hatreds virtually guaranteed continual chaos. Adding to pressures not of Fleming's making, Alexander Mackenzie, the Liberal Party leader, had promised the leaders in distant British Columbia a transcontinental Canadian railroad within ten years of confederation—on the threat of their opting out of the agreement and going it alone, or, worse, joining the United States.

The two founding political parties, the Liberals and the Conservatives, and their leaders, Mackenzie and John A. Macdonald, quarreled over every surveying decision Fleming rendered.

Mackenzie, a stonemason by trade, could, without apologies, be called flinty and gravelly; his constituents were the small farmers of Ontario. Macdonald enjoyed the support of the Montreal (English) business community, the establishment, and the crown-appointed governor-general. When not in his cups, he could be a charming and effective leader. Fleming, however, was not accustomed to second-guessing by any uninformed outsider, nor to tailoring his surveys to satisfy a pool of constituents and powerful supporters. To a politician, rewarding loyal backers with the promise of a rail line was instinctual behavior. What might look on paper as a modest expenditure became for Fleming a burden to be amortized over fifty years, requiring new surveys over untested soils and river crossings, and new crews and more provisions, especially in the face of a dependably long and harsh winter, that could run into millions of dollars. Approving political requests against engineering judgment was a violation of his professional commission as a civil engineer.

Lost, perhaps, in all the in-fighting was a simple human fact: Fleming alone had been asked to do in ten years what had taken scores of specialists nearly sixty-five years to accomplish in the United States. Canada's knowledge of its own far west in 1872, when the surveys finally got under way—particularly the mountain ranges of interior British Columbia and the most promising passes through them, as well as the costs and strategy for filling and bridging the swamps of the Canadian Shield—was hardly greater than President Thomas Jefferson's had been in 1803 when he appointed Lewis and Clark to survey the vast new lands of the Louisiana Purchase. Fleming undertook the surveys, camping out through summers and winters, consulting trappers and Indians, all the while taking notes on weather, water, soils, plants, and wildlife. He was surveyor, naturalist, scout, meteorologist and geologist, as well as the country's leading civil engineer.

Now, on his medical absence from Ottawa, the twin engineering assignments on the two rail projects were being handled by crew chiefs of varying competence but uniform ambition. Questions of his competence and even of his probity were being raised in Parliament, and his faithful friends and defenders—he was a man to engender deep loyalty—were finding themselves politically vulnerable. Fleming himself, although a personal friend of John A. Macdonald's (the irascible Mackenzie had few intimates), prided himself as a civil servant on never having voted, nor of ever having uttered a partisan comment in public. The effect of his aloofness was to render himself increasingly expendable.

The missed train, then, occurred when Fleming had the time and an unaccustomed release from the contentious minutiae of engineering to contemplate theoretical change, and it led within hours to an immediate solution—a twenty-four-hour clock, in which 5:35 P.M. would be transformed to 17:35h—but in the weeks to come it would lead to something much grander indeed. An idea for time zones and their relationship to longitude started forming. Four months later, back in Toronto, at a meeting of the now well-established Canadian Institute, he presented his first time paper on the reckoning of "terrestrial, non-local" time.

Heat still rises from the pages of that first paper as he remembered the night in Bandoran. "This was the first few days' experience of a visitor from a distant country to the United Kingdom," he wrote, "where untold wealth and talent have, during many years, been expended in establishing, developing, and perfecting the railway system!" His initial proposal was far too complicated, combining standard time zones with a single, universal time for the world, tying longitude to time and, of course, imposing the twenty-four-hour clock.

Its central difficulty, however, was its failure to endorse a prime meridian, a place where the time zones for the world were to begin. There is something almost perverse in his refusal to accept the obvious appeal of Greenwich, which he avoided on the grounds of arousing "national susceptibilities," a euphemism for French resistance. That resistance created obvious complications, which he addressed rather ingeniously. Instead of a prime meridian, he proposed an imaginary "chronometer" in the middle of the earth, a clock face like a giant gauge that would convert time to longitude and vice versa.

Fleming's chronometer resembles nothing so much as one of the earliest mathematical problems he had faced while studying for his surveyor's commission in 1848, a version of that insoluble puzzler, squaring the circle. The problem as stated: *Even the four sides of a quadrilateral figure inscribed in a circle.* In other words, can the area of a four-sided figure inscribed within a circle ever equal that of the circle? (Charles Piazzi Smyth, the astronomer-royal of Scotland, and one of the true characters in the history of the profession—"keeping as ever to his own orbit," a visitor to his observatory in Edinburgh once declared—devoted years to trying to prove the *pi*-qualities of the base of the Great Pyramid of Giza. The motive behind his mission was to disprove the utility of the godless French meter and to uphold the honor of the Imperial inch, which he contended had been invented and used by the Greeks. Needless to say, the effort nearly ruined his reputation as a serious scientist.) In devising the buried chronometer, Fleming was addressing a purely temporal version of the same puzzle—how to divide the universal day into local parts, without imposing an arbitrary prime meridian.

This is Fleming's intellectual habit: deductive, *a priori* reasoning, from general principle or universal theory, back to the particular, or individual, application. From soaring premise to the rubble of mere detail. In the nineteenth century both forms of reasoning, deductive and inductive, *a priori* and *a posteriori*, were considered legitimate and productive. Today we're more suspicious of the anticipatory leaps that deductive logic can sometimes make.

The paper on "Terrestrial Time" was far too difficult for commercial adoption in its day. For starters, it would require new dials to be glued over every watch- or clock face in existence (for which he made five U.S. patent applications, failing the test of uniqueness on four of the five insofar as innovative watch faces had been proposed, though never built, a generation earlier). There would be an outer wheel of twenty-four Roman numerals representing the time zones, and an inner wheel of twenty-four letters (omitting J and Z) representing the hours (plus a numerical innermost ring representing the minutes). It would be possible to read the time as M.22, for example, a mind-numbing exercise even today, but it confronts a problem we (or our children) might find intriguing. Why Greenwich, or any other meridian, we might ask? A land surveyor, or a ship's captain, might be wedded to exact geographic meridians, but what of virtual meridians in space? There's something inconsistent between our cesium-ion clock's ability to divide each second into billions of parts, and our toeing a line, a starting gate, drawn a hundred and fifty years ago.

Nature is uncharacteristically accommodating in matching a twenty-four-thousand-mile circumference of the earth to a ready-made twenty-four-hour day. Each of twenty-four time zones, therefore, occupies fifteen degrees, or approximately a thousand miles of latitude. It's obvious—but should rational people necessarily be expected to follow? In some ways, perhaps, we were taken in by a vast coincidence. We already know that Fleming saw time as a flowing continuity, and it is only natural that his instinct would be to honor the flow and not to dam it with an arbitrary, stop-and-go prime meridian. Is there a way of separating time from the constraints of a politicized, commercialized prime? Can time be freed from geography, history, and nationality all together? From that moment on, and especially after his second (simplified) paper in 1878, in which the buried clock was abandoned and a surprising new prime meridian was proposed, Fleming became the focus of time reform for the world.

The multiplicity of local times ("What times is it?") and their reconciliation to the single "cosmic" moment, was the motivation behind Fleming's first reforms. Ideally, he felt, everyone in the world should be on the same temporal page, we should know instantaneously, by a single mode of time-reckoning (like soldiers in the field or pilots in the sky), what time it is, everywhere. Time zones were appropriate for local considerations, he felt, but travel schedules and communications should be rendered in universal time.

Fleming's first proposal had a futuristic ring to it, rather like the "stardate" at the beginning of every *Star Trek* episode. It showed him to be rooted in practicality, and in his training as a surveyor (*"Point out how the longitude is formed"*), but open to abstraction. In an opening burst of enthusiasm, he even translated a railroad schedule into his new temporal language, not that he ever submitted it to the Time Convention of the Railroad Association. The North American railroad passenger would have to keep consulting, for another seven years, the temporal correspondences and competing schedules in Mr. William F. Allen's *The Railroad Traveler's Official Guide,* the American version of the Irish guide that had misled Mr. Fleming, and just hope that he had read it correctly and that no demons had crept into the text.

Telling time or keeping track of time outside of one's own township was an ongoing irritation in the 1870s, the degree of confusion directly proportional to the number of long-distance travelers using the system. The majority of people in the 1870s were still able to arrange time to suit their convenience, even arranging a noontime stroll to watch the dropping of the local time-ball and to take out their watches and make a precise adjustment. How better to maintain communal integrity? Every self-respecting town on the continent had a right to its own newspaper, its own baseball or cricket team, and its own individual time.

IN 1881 the venerable *Atlantic Monthly,* Boston's literary and intellectual voice, proposed a minor heresy: New England should adopt New York City time. Boston local time (or, to be more precise, that part of universal time that passed over the Boston meridian) was twelve minutes ahead of New York City's, enough to engender chaos in towns like Hartford, Connecticut, or in western Massachusetts, where the two standards clashed. The *Atlantic* was giving public voice, perhaps for the first time (and perhaps courtesy of a "leak" at the Railroad Time Convention) to the secret debates that had been animating the professional societies:

> Besides the argument of business expediency, there are other reasons for adopting the New York standard, or a practical equivalent counted from Greenwich, based upon the consideration of what is best for the whole country. There has been a steadily growing public opinion in favor of dividing the whole of the United States into five sections, such that the time of one section shall differ from that of the preceeding or following section by a whole hour, so that the minutes of time shall be the same from Portland to San Francisco, and the local time in any will not differ more than half an hour from the standard time adopted. . . .
>
> By calling the different sectional times by easily remembered names, beginning with Newfoundland, and calling the time used over Newfoundland, New Brunswick, and Nova Scotia Eastern time, we should then have Eastern, Atlantic, Valley, Mountain, and Pacific time,—this last comprising the Pacific slope, British Columbia, and Vancouver's Island.

Buried beneath those very reasonable sentiments was another reality: Boston, no longer the pacesetter of American opinion, could retain its influence only in consort with other East Coast cities. The dynamism of the country lay in newer cities to the south and west, where railroads could fan out in every direction, serving their markets and importing their needs. The old seaports were constricted. "The adoption of such a standard [i.e., a single "Atlantic" time, which we now call Eastern] would free the matter from any objection based upon pride in keeping to a more local time, and would enable the cities of Boston, New York, Philadelphia, and Baltimore to have a common time for all business purposes." It was time for the old America to consolidate its strengths and overcome its narcissistic differences.

Gradually, even without industry innovation, the varieties of local times began to coalesce out of sheer economic necessity around the larger trading centers. Connecticut abandoned Boston time (and its pretense to a New England identity) in favor of New York City time across the entire state. A time-defined "Chicagoland" began to emerge in the tristate area around that metropolitan center. The number of official times declined to a hundred, then eighty, and finally, to forty-four, where the momentum stalled. Pressure was building, but no further progress, at least in public, was being made.

By 1880, England had been on standard time for over thirty years. Why couldn't Americans simply drop their local attachments and adopt a single standard? A number of theorists recommended endorsing God's handiwork and splitting the continent right along the Mississippi valley, into two time zones. Time reform in Britain had also started with the railways, when the Great Western unilaterally dropped all local times from its schedule, unifying its routes to the time signal of the Greenwich Observatory. Other lines had been forced to follow suit, and in 1852 Greenwich was recognized as the Parliamentary standard. Britain enjoyed the benefits of standardization before any other industrialized nation, and the results—technologically, commercially, even culturally—were astounding. So why not America? For America, it was a question of scale.

Europe was scarcely the test for standard time. To Americans, England seemed a cramped and narrow place: from easternmost East Anglia to westernmost Cornwall, the solar difference was about equal to that between the American cities of Boston and Pittsburgh. The east-west span of the entire British Isles, in fact, from East Anglia all the way to the western shore of Ireland, does not match the distance between New York and Chicago. When hours, rather than minutes, separate the population centers, practical as well as political considerations override the simple expedient of an act of Parliament.

Considering the impossibility of reaching universal consensus today on nearly any scientific or political protocol—whether it's rain-forest preservation, global warming, population control, landmine removal, genetic engineering, or nuclear testing—the fact that implementing standard time for the world was finally agreed to in civilized debate, did not cost a penny, nor the loss of a single life, speaks to a spirit of cooperation, and a quality of leadership, that we may never see again.

STANDARD TIME for the world began in the creative use of a technical misadventure, and it took another turn, not necessarily for the better, two years later, when Fleming was invited to deliver his second, simplified, time paper to the 1878 meeting of the British Association for the Advancement of Science, held in Dublin. His paper had caused a mild sensation at the Canadian Institute on its first reading, causing the governor-general of Canada, the Marquess of Lorne (Queen Victoria's dandyish son-in-law), to forward copies to Sir George Airy, the astronomer-royal, and to the Colonial Office in London for purposes of having it translated and sent to the world's leading astronomers. Airy relayed his recommendations directly to the Privy Council. Fleming's paper would not, however, enjoy a sympathetic hearing at the Dublin meeting. It would receive no hearing at all.

It's never easy, coming from the colonies, to storm the imperial center, lacking the proper degrees and credentials, but one should not have to suffer additional insults on their behalf, as

Airy's recommendation would soon impose. It was the beginning of the bleakest year and a half of Fleming's life. The indecisiveness of his engineering commission and attacks from his parliamentary enemies combined, within months, to force his dismissal from the Canadian Pacific. One can feel enormous sympathy for Fleming's earnestness, his belief in the British system, and for the multifarious vanities of the Victorian public thinker. He was not a vindictive man, but in a very long life he never forgot, nor forgave, the wounding he took in those two weeks in Dublin, waiting to deliver his paper. For the remaining thirty-five years of his life, he would be, essentially, honored but unemployed.

He wrote to the secretary of the conference from his Dublin hotel room, on the last day of the session:

I reside in Canada and can only enjoy the advantages of the association by travelling a long distance at considerable inconvenience. I determined to be present at Dublin. I prepared a paper which I thought clearly came within the objects of the association as set forth. I complied strictly with the rules as laid down under the heading notice to contributors. I received a proper acknowledgement before the end of July and was then informed that my paper be brought before Section A.

I arrived in Dublin on the 14th inst. bringing with me instruments and drawings prepared at considerable trouble and expense and addressed a note to the secretary of Section A and informed him I was ready to read my paper whenever convenient. On the morning of the 15th I called and was told that I would be informed in due course.

Receiving no reply, on the 17th I again called. I was then informed that the Convention had decided to have my paper read on the 21st (today).

On the 20th I received notice that my paper was placed on the list for that day (yesterday) and on examining the list I found it at the end, there being about a dozen papers in all to be read before my one could be heard. I attended the section until the meeting closed but no opportunity of reading my paper was afforded me and on enquiry I was told that the section would not again meet.

On the list of papers to be read today no mention is made of my paper, and I understood this is the last day of the session.

I do not propose to allude to the paper I intended reading beyond saying that I was most anxious to bring the question with which it deals before the public through the British Association. I submit merely the bare facts connected with a fruitless effort on two occasions and point out to you how difficult it has proved for a total stranger to get a hearing. I shall not further trouble you with any comments, but I must express my very great regret that I have felt it necessary to make you aware of my feelings.

In general, when angry, Fleming cultivated an air of pique rather than confrontation. The Dublin snub, however, was the sort of wounding that he could not forgive or forget. Allusions to it appear in all his writings, whenever the question of standard time comes up. It had been a terrible two years: the parliamentary inquiries, newspaper criticism, satirical cartoons, the loss of the CPR commission. As Hemingway put it in "My Old Man," "Seems like once they get started, they don't leave a guy nothing."

He never openly speculated on the reason for his deselection from the session, but there is evidence that Airy himself had intervened against his appearance. Airy, distinguished but irascible, was the very man who had introduced standard time to Britain nearly thirty years earlier. Whatever his reasons—age (he was seventy-eight), jealousy, or perhaps even a fear that the observatory might lose its profitable monopoly on the selling of Greenwich charts—he ridiculed Fleming's various papers up to the time of his retirement two years later:

I set not the slightest value on the remarks extending through the early parts of Mr Fleming's paper. Secondly, as to the need of a Prime Meridian, no practical man ever wants such a thing. If a Prime Meridian were to be adopted, it must be that of Greenwich, for the navigation of almost the whole world depends on calculations founded on that of Greenwich. Nearly all navigation is based on the Nautical Almanac, which is based on Greenwich observations and referred to Greenwich Meridian, and the number of Nautical Almanacs sold annually exceeds, I believe, 32,000. But I, as Superintendent of the Greenwich Observatory, entirely repudiate the idea of founding any claim on this. Let Greenwich do her best to maintain her high position in administering to the longitude of the world, and Nautical Almanacs do their best, and we will unite our effort without special acclaim to the fictitious honour of a Prime Meridian.

Airy's conclusion strikes the scornful, above-the-battle stance to which astronomy often aspires. (One need only recall the arguments recorded in Dava Sobel's *Longitude*, the contempt of an earlier astronomer-royal, Sir Nevil Meskalyne, for the provincial clock-maker John Harrison.) Airy's recommendation to the Privy Council was to abstain from any "novelty" or "social usage," on the principle that government intervention might prove more harmful than the recognized inconveniences enumerated in Fleming's paper. He closed on a note of sheer condescension, suggesting that Fleming and the Canadian Institute would do better to petition the Dock Trustees of London, Liverpool, and Glasgow.

Why such an attitude from the man who, in 1852, had introduced the daily dropping of a public time-ball tied to the Greenwich signal? It was Airy who had brought standard time and all of its advantages to Britain. Was he merely protecting his place in history, jealous of a younger man, a foreigner with a broader but less scientific vision? Or was it the reluctance that comes with long tenure to engage or endorse a possible fad? Or something more calculating, fear that Greenwich under Fleming's plan might lose its profitable preeminence in the world? Britain, it should be remembered, thanks largely to Airy himself, was unaffected by the dangers and dislocations caused by the rampant time standards of North America and the rest of the world. They were above it all.

And there is an even more convincing argument. Fleming's 1878 paper had proposed a prime meridian, but it did not run through Greenwich. The Colonial Office was only too happy to withdraw its support or, indeed, to be relieved of having to take any position. Airy's successor, a supporter of standard time, could not provide a motive for Airy's unfortunate behavior.

6

The Practice of Time

Time was in the air.

—E.T.D. MYERS, former president,
American Railroad Association (1904)

FIFTY YEARS AGO people of modest means used telephones only for local calls. Awful news arrived by telegram.
"Wires" communicated a palpable respect for time and space.
Distance meant dread, delivered in a yellow envelope. The farther the distance, the greater the certainty, the deeper the tragedy. Return messages took agonizing hours. I remember standing
under tall tables in small-town Western Union offices in the Deep
South, watching my mother edit out words or try new combina-
tions that might save a dime, or pare a message down to its purest
feeling.

When I was seven years old, in Leesburg, Florida, in 1947, our
next-door neighbor, known only as Big Mama to the children in
the neighborhood who gathered nightly on her porch, told us
stories of her adolescence and marriage—*before* the Civil War.
Here was Time embodied, in the traditional Southern manner,
through tales and a survivor's voice. It was my moment, like Dan
on Pook's Hill.

After Big Mama, Faulkner's voice, when I got to college, seemed almost neighborly. And long after college, I came across Walt Whitman's "The Night I heard the Learn'd Astronomer":

> *I wander'd off by myself*
> *in the mystical moist night-air, and from* time to time
> *Look'd up in perfect silence at the stars.* [Emphasis added.]

How Whitmanesque—that the starlight falling on earth (*on him!*) that night was as young as the evening and ancient beyond calculation. It was his privilege to wander between the new and the ancient. What a perfect example of a "new" American voice, a voice for the rails, the goldfields, the war, free of New England, free of constraint, unafraid to celebrate the rawness of his vision. And now I can't help wondering who, exactly, that "learn'd astronomer" might have been. Perhaps it was Sir John Herschel, England's grand old man of astronomy, the first scientist in the nineteenth century (way back in 1828) to propose a reform of the equinoctial day. Or the young Simon Newcomb, the Canadian-born scourge of standard time, long-standing enemy of Fleming, and, eventually, head of the U.S. Naval Observatory.

The poets' broadest intuitions are confirmed by modern science. We *are* eternal. Our bodies are the skeletons of dead stars and in the fullness of time we return to our makers. The universe seems a vast Manichean metaphor—the super-hot, super-dense quantum cloud, timeless and spaceless, of infinite heat and density that preceded the big bang, twinned to the universal collapse of light and matter into the infinite gravity of the black hole. From one, the universe and all its creations, including time, are released. From the other, nothing, not even time, escapes. Both are mathematically inexpressible "singularities," predicted, but even resisted, by Einstein. In the Alpha and Omega of creation, contrary to his assurance, God *does* play dice with the universe.

Seven years after those Florida nights, in the mid-1950s, my parents moved north to Pittsburgh (as it is now spelled). Since they both worked, often till midnight, I found myself going to movies three nights a week during my high school years. Movies ran continuously, like New York subways. We entered, we sat, we munched our popcorn, and we watched the movie. The moment we sat, no matter how deep into the plot it might have been, became our beginning, our prime meridian. We stayed through "The End," but the movie's "end" was just another meridian; we still had half a movie to go before *our* movie, our "day," could end. We stayed through the cartoon, the news, the coming attractions, which for us were not introductions nor trailers but an interruption of the feature. For us, the opening credits came in the middle. We already knew the ending—the suspense was learning how it started. Everyone who entered as we did fashioned a different movie; they left when *their* beginning came round again, making it their end. And so, everyone in the theater watched the same movie, but also a different one, based on one's private beginning or, in terms of this book, one's own prime meridian. Because each of us had different plot preparation and different depths of reference, some of the audience laughed when others were sober. Some of us snickered at tender or suspenseful moments.

Was there one movie, a dozen, or a possible infinity? That's what time had been like before the Prime Meridian Conference.

THERE CAN be only one sunrise and one sunset per day. The rotation of the earth is "natural," God-given, but there is no limit in theory (and sometimes in practice) to the number of hours that make up that day, or how many minutes we choose to call an hour. They are "rational," man-made. There is nothing God-given in the length of the second, minute, or hour. Traditional Japanese timekeeping had employed flexible hours, expanding in the summer, shortening in the winter, to keep pace with the sunrise and sunset. The French Revolution in its zeal to create a new consciousness, dictated hundred-minute hours, a twenty-hour day, the ten-day week, and twelve thirty-day months per year. Despite the appeal of decimalization, however, not even Robespierre could change the simple fact imposed by nature that each "day" marks one rotation of the earth, and that the year marks one complete orbit around the sun. And even the French Revolution could not suppress the fact that every longitude on earth experiences the sunrise at a different local time. In recognition of the French, however, it should be admitted that even though the Revolution failed to rewire the human consciousness, France never really abandoned—nor did it fail in—its overpowering ambition. In the modern world, the French mania for universal order has been rewarded. We don't observe hundred-minute hours, but we do observe universal time, such as the airlines' Zulu time, regulated by a signal originating in Paris.

So why do we observe the Greenwich meridian, and not that of Yokohama, New York, or Buenos Aires? This was the question faced by Sandford Fleming in 1878, at the time of his second time proposal. He had abandoned the buried chronometer of his first paper, the complicated watch-faces and the interchangeability of time and longitude. He clung only to the notion of a universal day, the twenty-four time zones and his signature twenty-four-hour clock. The popular, or democratic, answer to the question of Greenwich is that most of the world's shipping employed Greenwich charts, which made Greenwich the obvious choice on the basis of convenience alone. So why the delay? Fleming was alert to the fact and sensitive to the feelings behind it, that the election of a British prime was sure to arouse national enmities, especially from the proud tradition that would be forced to surrender its *ligne sacrée,* the Paris meridian, nine minutes and twenty-two seconds ahead of Greenwich. The French position, which Fleming supported (up to a point) was that there should be scientific, not merely commercial, reasons for choosing an international standard.

There were overwhelming historical, economic, and political reasons for choosing Greenwich, but *not* necessarily any compelling astronomical reason; that is, it bore no scientific rationale. No single longitudinal meridian is scientifically superior to any other, and astronomers pride themselves on a tradition of lofty disinterest in mundane affairs. Many astronomers, in fact, opposed the entire standard-time movement on the grounds that it might possibly involve them in matters they considered merely political—that is to say, unworthy. All north-south meridians are equal; it's the east-west meridians, the latitudes, that are scientifically predetermined. No one would designate the fortieth degree of latitude as an equator, even if the majority of the world's population lived along it. There can be only one equator, but every longitude on earth, every great arc that passes over the two poles, revolves about the sun once a day. Paris or Washington or Yokohama are no different from any place else, including the empty Pacific, so why should Greenwich be favored above any other?

The observatory at Greenwich was indeed honored and renowned (and profitable, due to the selling of its charts, the so-called ephemerides, to the 90 percent of the world's shipping that employed them), but the national observatories in Rome, Paris, and Berlin, and the Naval Observatory in Washington were no less well-equipped. Without science to give backing to a particular meridian, there could be no "neutral" determination, and perfect scientific neutrality, not commercial popularity, was the nonnegotiable French demand before they would consider joining any international convention. That was the great challenge Sandford Fleming faced in his subsequent papers: to use the political and commercial advantages of Greenwich, yet appear *not* to use them at the same time. The drama behind the election of Greenwich, against fierce opposition as well as the rejection of a sophisticated compromise, is the major part of the struggle for standard time. The decision would not be rendered until 1884, at the Prime Meridian Conference.

FOR THE fifteen years between the closing of the American frontier, in 1869, and the conference, a series of proposals for reducing the number of local times was launched and debated. In 1872 Professor Charles Dowd, principal of Temple Grove Ladies' Seminary in Saratoga Springs, New York (now Skidmore College), revised his original proposal, which had been based on the Washington meridian, and floated a five-time-zone system for North American railroads that is nearly identical to the system used today. His time zones varied each by one hour, each zone covering fifteen degrees of longitude, counting in fifteen-degree leaps westward from Greenwich. Professor Benjamin Peirce of Yale University, as the *Atlantic Monthly* had written, had also proposed a similar reform, as had Professor Abbe. Time was in the air, but all of the reforms thus far were attached to railroad use only, confined to North America, and dependent upon the still-unratified Greenwich meridian.

Dowd, Abbe, and Peirce all appeared on the verge of proposing standard time for the world, and not just for North American railroads and their passengers. But no one could imagine the mighty railroads changing their time-standards merely by crossing an arbitrary, invisible line. Railroads could change time, but time could not change railroads. By extending the fifteen-degrees-per-hour series of time zones around the world, looking beyond the Atlantic and Pacific shores of North America, *et voilà,* they would have it. But there are two very good reasons why their solutions would not have worked in 1874, and why, in fact, it would take another ten years to overcome them.

First of all, and most obviously, the most progressive American reforms were predicted on Greenwich, but Greenwich had not been agreed to by the world. The United States Navy, and its commercial fleet, and of course Britain and its colonies all used Greenwich charts (the two countries alone accounted for nearly the entirety of Greenwich's popularity), but ten percent of the world did not. There were, in fact, ten official prime meridians in use at the time, all of them historically justified, all with their national pride and clientele intact, which made shipping schedules nearly as confusing as catching trains. The United States was free to adopt whatever standard time it wished, of course, but it would have no relevance outside of its own territory, or the industry it was meant to serve. And, second, as extensive as North America was, it still did not have to cope with the change of dates, a date line, and where such a line might be drawn.

Dowd's 1872 innovation had been to break with American practice by dropping the Washington meridian and adopting, or simply assuming, the Greenwich prime. Dowd, a Yale-trained professor (voted outstanding member of the class of 1857), had been provoked into action originally by those three "official" railroad clocks in the Buffalo train station reflecting Albany, Columbus, and Buffalo time. He had written, with philosophical exasperation: "The traveler's watch was to him but a delusion; clocks at stations staring each other in the face defiant of harmony either with one another or with surrounding local time and all wildly at variance with the traveler's watch, baffled all intelligent interpretation." But Professor Dowd was easily dismissed as an impractical dreamer (a common designation for professors at most times in American history, and never more so than in the Gilded Age), and an outsider to the railroad fraternity. His proposals were politely listened to by William F. Allen, secretary of the American Railroad Association, and other grand-trunk managers, and shelved.

The revolution in time-reckoning on the North American continent came not through theory or governmental intervention, but by way of old-fashioned American commercial innovation. On April 8, 1883, in St. Louis, at the semiannual meeting of the General Time Convention of the American Railroad Association, fifty managers of grand-trunk railroads voted to accept a plan put forth by their secretary, Mr. Allen. It would reduce the number of time standards in use by American railroads from nearly fifty to a mere four. Allen named them Eastern, Central, Mountain, and Pacific, the same four names, though not the same four zones, that are used today. A fifth, that of the Canadian maritime provinces, designated Intercolonial, was added a few months later.

If ever a demonstration of temporal confusion were needed, St. Louis was the city to provide it. Delegates arriving on any of fourteen different railroads would experience the most blatant example of temporal meltdown to be found anywhere in the country. St. Louis observed six official railroad times. Although Allen's reforms were presented to the managers as a simple business decision, history has shown that the standardization of railroad time across the continent would have fundamental implications for every part of American society. A contemporary journal hailed its implementation as "one of the most complete scientific successes of the century," and "the first step in an inevitable process of world-time standardization." Allen's formula for railroad standardization in North America, however, played no role at all in the world standardization movement, partially for reasons already mentioned, and even more centrally, for the manner in which Mr. Allen designed it. The prediction of world standardization proved accurate, however, thanks mainly to the efforts of Fleming and his Washington friend, Cleveland Abbe, eleven months later.

And as for the first part of that glowing magazine quote, a claim of such scientific success deserves closer scrutiny: Darwin, Pasteur, Edison, Bell . . . and William F. Allen?

Allen was a trim, thirty-seven-year-old, New Jersey–born civil engineer. His father, also a civil engineer and an army officer, had been killed in the Civil War the same year Allen, at sixteen, began work as a rodman on the Camden & Amboy. Six years later he was appointed resident engineer on the West Jersey. The manager of that line, General W. J. Sewell, was later elected to the Senate from New Jersey, and remained an important patron. Allen's rise was a familiar pattern in that frantic, brawling, post-war, frontier-closing, land-grabbing era. Call it the Ben Franklin syndrome—a bright, self-educated, ambitious near-orphan works hard, rises early, stays late, attracts rewards and influential friends. Success came early, and conspicuously.

Allen's temperament, "the genius of hard work," his experience and contacts, combined to make him a formidable advocate for, and often a bulwark against, new ideas. He understood the tight fraternal nuances of railroad management and operated as the buffer between an increasingly impatient public, outraged by the dozens of competing and often unpredictable time standards, and a heedless, immensely profitable industry. Railroads were the driving force of the economy and a magnet for ambitious freebooters. The great fear of the industry was that their profits, competitive practices, and monopolistic ambitions—along with mounting complaints from manufacturers, farmers, and passengers—would attract government regulation. The frequently drawn analogy of nineteenth-century railroads to the contemporary world of computer entrepreneurs and dot-com cowboys is not misapplied. The arrogance of such industries can be annoying, but the real fear is that behemoth technologies, like the railroad, or the computer, simply obliterate earlier modes of transport or communication, leaving the public helpless, as in the case of airline passengers trapped by a snowstorm, with nowhere to go and no other way to get there. The technologies they replaced quickly became relics, museum pieces.

Over the Decade of Time, in his official capacities as secretary of the Association, as editor of *The Railroad Traveler's Official Guide* (the rail passenger's indispensable machete through the temporal jungle), and as a member of the American Metrological Association, Allen had reviewed dozens of proposals for time reform. Not all of them were flawed, but every one had been rejected. Those originating from Professor Dowd, despite their cogent argumentation, detailed maps, and painstaking, station-by-station research, had been listened to tolerantly and approvingly by the railroad managers. When they were originally proposed, their solutions were deemed premature. Meltdown loomed, but had not yet occurred.

Dowd's proposed zones were neat, clean, and geometrical. They followed longitudinal meridians with a serene indifference to political and commercial boundaries in their path. Every fifteen degrees, starting at Greenwich, marked the center of a new hour, retaining the same minute and second. Allen's zones, by contrast, were as fussy as a Victorian parlor. They were the result of a compromise between railroad intransigence and passenger frustration. Yet, at the end of the day, Allen was the railroad man, and his zones were more accommodating to railroad authority than to passenger convenience.

To assure his audience, he stated: "It has been my earnest endeavor to look upon this question purely with a view to the practical requirements of the railways, to pose nothing which is unprecedented, or that will bring about a condition of affairs which, viewed in its most unfavorable light as affecting the running trains, does not at present exist, or has not been already practically overcome; to avoid making the remedy worse than the disease."

In other words, no ground-shifting proposals. He also articulated a position that can be seen as far-reaching, at least insofar as it reflects the view of the primacy of corporate authority over political jurisdiction: "From a railroad standpoint we have nothing to do with state lines or national boundaries, but must confine ourselves purely to the needs and be governed by the limitations of railway operations." Here, Allen was running a risk of attracting government scrutiny, even in midst of the friendly, laissez-faire style of postwar Republican presidents, none more typical than the sitting Chester A. Arthur.

In Allen's time the public were accustomed to city, county, and state governments deciding, or at least debating, proposed changes that might have profound effects on their lives. In the case of railroad standardization, a purely "corporate" decision with enormous public consequences had been made in private, without citizen input or oversight. A fundamental reordering of every citizen's private day was about to occur, ordered for the convenience of an interstate industry that took for granted its jurisdiction over local concerns. The lines between corporate freedom and governmental oversight were obviously hazy. Allen immediately began lobbying route managers and regional presidents. The proposals went into effect seven months later, as soon as the entire Association read them and approved.

We are not scientists dealing with abstractions, but practical business men seeking to achieve a practical result. We have a common language, a common standard of money, of weights and of measures, notwithstanding the enormous extent of our country; but even approximate or relative common time is yet to be achieved. Less than ten years ago we had seventy standards, today we have about fifty. For this much relief let us be duly thankful.

Allen's revisions also favored five time zones for North America, and acknowledged the Greenwich meridian; but, unlike Dowd, he'd arrived at his maps by consulting the railroad grid and noting where the preponderance of similar time standards were clustered. Thus, he constructed his own zones, reflecting as faithfully as possible the preexisting railroad operating standards so as not to inconvenience published schedules. The edges of Allen's time zones were necessarily ragged, as they followed the existing rail lines, allowing them to retain their time standards to their terminals, or to the next major rail junction. (In other words, they did not automatically change upon crossing an invisible meridian.) He was lenient with a number of major lines, allowing them to operate outside the stricter standards of their neighbors. Yet although Allen's zones were pushed a little to the east of Dowd's, in practical application the effects were very similar. The differences were more in the biographies of their devisers than in the outlines of their respective time zones.

Allen's eastern zone, for example, ran from Maine to Detroit, then south to Bristol, Tennessee. *But* . . . railroads in Ohio and Pennsylvania west of Pittsburgh, and all those in Georgia, would be included in the western (i.e., central) section. Five other major railroads would be allowed to run to their eastern terminals in Buffalo, Charlotte, and Salamanca, New York, by the central time standard. Allen's bold proposal was beginning to look nearly as complicated as the original problem; the remedy *was* nearly as bad as the problem.

One solution that did not occur to Allen (and which, admittedly, courts absurdity and takes some getting used to, even today), is the simple expedient of printing all schedules in local time. A train might leave an eastern city at noon, and arrive in a nearby central-zone station *before* noon, as we often do today when flying. Allen's plan retained the train's original time standard for as long as possible, all the way to its terminal or point of transfer. Dozens of exceptions were permitted for popular routes. A complicated map accompanied the new reforms.

Dowd and Allen embodied some of the great philosophical divides of the nineteenth century. Allen counted trees; Dowd saw the forest. Allen was a tinkerer, Dowd a synthesizer. Allen's plan was derived from railroad usage; Dowd's from passenger convenience. Allen was the pragmatic, politically expedient man of the world, Dowd the dreamer, the impractical professor. In the autodidact world of nineteenth-century American science and philosophy, deduction (*a priori* thinking) and induction (*a posteriori* thinking) were both acceptable approaches to a problem. And, I might add, Dowd was a New Englander, a child of transcendentalism; Allen was the New American Man.

On Sunday morning, November 18, 1883, railroad standard time ("Vanderbilt time," sneered its opponents, but it was more accurately Allen time) became a reality across North America. That Sunday came to be known as "the Sunday of Two Noons," since towns along the eastern edges of the four new American "time belts" had to turn their clocks back half an hour, creating a second noon, in order to conform to towns along the western edges of the same belt. No one in the country would "gain" or "lose" more than half an hour of his life. The dominant technology of the age had set the new time standard. Up in Ottawa, Sandford Fleming hailed it as "a quiet revolution."

Within days, about 70 percent of schools, courts, and local governments had adopted railroad time as their official standard. The federal government, much to Allen's relief, did not get on the bandwagon. (In fact, Congress did not get around to ratifying standard time until 1918.) For the first time in history, Boston and Buffalo, Washington and New York, Atlanta and Columbus, San Francisco and Spokane, all shared the same hour and minute. It didn't matter that Boston would be bright with the new day while Wheeling was still dark. In fact, it didn't matter what the sun proclaimed at all. "Natural time" was dead. Some towns, like Bangor, Maine, and Savannah, Georgia, refused, out of religious faith or plain old stubbornness, to go along. A city like Detroit, perched at the edge of eastern and central times, could not make up its mind, and voted itself, over the course of many years, in and out of both times before settling on its "eastern" designation. For many years people in Detroit would still have to ascertain, in setting appointments, "Is that solar, train, or city time?"

The 1884 Prime Meridian Conference that set standard time for the world followed the Sunday of Two Noons by less than a year, but the model of standard time with which Allen is associated, that of railroad standardization, played no role. The conference protocols were based on Dowd-like proposals, although Dowd was not named, nor even invited to the conference.

IN 1904, Allen published a small book entitled *Standard Time in North America, 1883–1903*. The first nineteen pages recapitulate his central role in the standard-time movement (with brief and dismissive mentions of Fleming and Dowd), while the final sixty pages are letters solicited by him from his railroad colleagues, conferring upon him undisputed credit for devising and implementing standard time. "Of all living men you are the one entitled to the credit of inaugurating the system of Standard Time," wrote one. "I never heard of Mr Chas. F. Dowd in connection with our Standard Time; you were the only person known in the matter to me," affirmed another. Only Mr. E.T.D. Myers, it seems to me, struck a note both just and historic: "Time was in the air." It was "in the air," he goes on to say, the way freedom had been in the air a hundred years earlier. Mr. Thomas Jefferson got the credit for *that* little revolution, but (like Allen) he was propelled in his course by the tides of history, and by able assistants.

Most of *this* book is an application of Myers's modest but far-reaching insight. Time was in the air in every human endeavor. The standardization of time, the overthrow of "natural" time, were necessary preconditions to scientific, technological, and artistic innovation and experimentation, and they all came together in the decade of time.

Allen was on surer ground in claiming that the adoption of standard time had a cascading effect on other forms of standardization, among them track gauges, uniform couplers, safety standards, freight rates, and wage scales. Until standardization, the Baltimore & Ohio, America's oldest railroad, had employed a seemingly sanctified track gauge of four feet, eight and a half inches ever since its founding in 1830. Most railroads, with the exception of the Erie, followed the B&O. That width, however, had been the standard separation of stagecoach wheels, a standard that, in turn, commemorated the distance between the centers of stone tracks on Roman roads built to accommodate the axle length of military chariots. And so goes the process of natural thought, unchanged for two thousand years. Thus do Chinese iron plows come to America.

Allen was an engineer, not a philosopher, but his claims are philosophically germane to the larger concerns of this book. Standardization of time was a sophisticated abstraction that actually improved practical commerce, and lent communications "real time" coordination. Standardization of time is part of an ever larger shift of consciousness, toward secular rationalism, and away from "natural" authority. And standardization was not solely a reflection of railroad convenience. In his capacity as America's chief weather forecaster, Cleveland Abbe received dozens of hourly, data-bearing weather transmissions from dozens of offices, hundreds and even thousands of miles away. In the local-time world of the 1870s and early eighties, he had to translate each of those separate local times of transmission into a single "real time" in order to track the movement of storm fronts and weather patterns. Without standard time, his "probabilities," as he insisted on calling his forecasts, would have had as much accuracy as the *Old Farmer's Almanac.*

William Allen, Charles Dowd, and Sandford Fleming are the acknowledged fathers of standard time, but only Fleming spoke to the world. Allen's role is today celebrated with a gleaming bronze plaque in the remodeled grand concourse of Washington's Union Station that remembers him as the "devisor and implementer" of standard time for North American railroads. Sir Sandford Fleming boasts a historical marker on a knoll outside the public library in his native Kirkcaldy, Scotland, where he is memorialized as "the inventor of standard time for the world."

Charles Dowd, the tragic figure among the three, died in a rail-road-crossing accident in Saratoga Springs in 1904, and was memorialized with a bronze plaque in the First Presbyterian Church of Saratoga Springs for his role as educator and inventor of standard time. Fire razed the church in 1976. Only a few unmelted fragments of his memorial were saved.

Part Two

TIME WAS IN THE AIR

7

Notes on Time and Victorian Science

THE SPEECHES and papers on the Lake Ontario beaches and the formation of Toronto Harbour that Fleming delivered to scant audiences at the Canadian Institute in 1850 are modest examples of the noble tradition of passionate amateurism in the sciences that fairly defined the intellectual climate of the first half of the nineteenth century. Natural science was an intellectual seducer, undertaught and unrewarded, feeding the speculative frontier. Jules Verne's undersea and lunar adventures were not understood as pure fantasy. They, and Percival Lowell's "discovery" of the canals on Mars, respected a prevailing popular attitude. Even into late century, many educated people assumed the connectedness of life on the surface of the earth with extraterrestrial civilizations, as well as with undersea and even so-far undiscovered life in temperate, subterrestrial pockets. The idea of a habitable solar system was a Victorian commonplace. (Following the "refutation" of the canals, it has taken another hundred years to restore a similar respectability to the idea of an inhabited universe. It's the mark of our humanity. Many astronomers today project habitable zones, vast liquid oceans, on Jupiter's moons.)

Natural science in the era of "natural time," before the revolutionary period of rationalization that began in the 1850s, attracted the wayward eccentrics, the mathematically gifted or the profoundly curious, from approved professions like theology, engineering, and medicine. Botany, geology, astronomy, archaeology, mythology, and linguistics were all enriched by a colorful assortment of autodidacts and hobbyists. Even the greatest among them, Charles Darwin, was self-taught, although the thirty years he spent contemplating the results of his youthful voyage to the Galápagos virtually created the condition of modern science, the hinge between the natural and the rational worlds, between amateur and professional science.

Amateurism in its core sense—"for love"—had not yet deteriorated into dilettantism, as it would from the 1860s (thanks largely to the higher standards and demonstrated mastery of *On the Origin of Species*). Amateur scientists were often clerics, or sometimes barristers, who were both, after all, the most broadly educated members of their societies, for whom science and religion, or science and society, were obviously entwined. "Natural theology" was a comforting doctrine that did not conflict with the close scrutiny of nature, nor with the search for laws that governed its behavior. Nature was the face of God, and laws governing it were codices to God's plan. To know nature was the most profound way of intuiting God. Theologians, like Fleming's friend the Reverend George Grant, could devote six days of the week to the study of God's handiwork and the seventh in writing a sermon to praise it.

They kept journals, they sketched, mapped, and painted. On Fleming's ocean-to-ocean survey of 1872, the party included not only its historian, George Grant, but an amateur naturalist—always the first to rise, to climb the cliffs, to investigate the damp rotten logs, to scrape the underside of tree bark for beetle grubs, and gather the prairie flowers. The party was often obliged to wait, not just for rain and blizzards and buffalo herds, but for the return of the naturalist from an unauthorized side trip.

The "rambling" amateurs were prospectors of the new and interesting: flowers, fungi, insects, birds. They trained themselves to be accountable. In the decade of the 1850s especially, as we know from Thoreau and Dickens and Whitman, among many others, man's relationship to nature was changing from one of reverence (and a certain amount of fear) to one of protection. With industrialism irreversible, nature took on an added value of nostalgia. Nature no longer ruled in Western culture, but the rules that governed it offered themselves as the source of all undiscovered knowledge. Like Fleming, that *a priori* thinker, or like Cleveland Abbe, who discarded a career in academic astronomy in order to become a meteorologist, the passionate amateurs, especially in the Anglo-American tradition, were often impatient with lab work and experimentation. Abbe begged his parents to desist from finding him a prestigious teaching post; he didn't want to be stuck in a classroom, parroting lessons to restless boys. He wanted to be of practical use to humanity. They were capable of great anticipatory leaps that today would win no support from granting agencies. Their major ideas were already implanted; now they were looking for supporting evidence.

Victorian geologists clustered around the hillside cuts of new railbeds, where they found evidence of ancient seas and fossils of marine creatures hundreds of miles from any modern, warmwater ocean. Coal seams revealed ancient marine habitats. Languages, living and dead, were linked on the basis of obscure syntactical similarities by linguistically trained missionaries and colonial officers. Minute fossil evidence established the chain of evolution. For every Charles Piazzi Smyth, the anti-meter zealot determined to force evidence into the mold of theory, there were dozens making their quiet contributions. The same issue of the *Canadian Journal* (the Institute's publication) that contained Fleming's Toronto Harbour maps and findings on its history and preservation boasts a lawyer's notes on meteors, a minister's remarks on early Roman history, and another barrister's comments on "new Genera and Species of Cystidea from the Trenton Limestone." A Dr. Craigie and his son made an exhaustive list of indigenous plants found in the neighborhood of Hamilton, Ontario, and their dates of first flowering. (It would be interesting today, in light of global warming, to compare those dates of first flowering with current dates of flowering, if they can be established.)

During the construction of the Intercolonial Railroad, from 1867 to 1872, Fleming had survived Parliamentary inquiries about extravagance (and possible collusion) for ordering iron bridges instead of wooden. Wooden bridges made financial (and "natural") sense in the maritime forests, being infinitely faster and less expensive to build. He defended his choice on the rational basis of permanence and fire safety in that cinder-spewing age. Heavier bridge pilings, however, required deeper excavations and more careful soil analysis. Maritime river clays—and not Ontario sands—posed hazards to bridge pilings, in some cases signaling falsely that bedrock had been reached. In this case, unlike on the CPR, his engineering caution, and its attendant delays and expenses, were totally borne out. There was not yet a science, or a name, for some of the discoveries Fleming and other early engineers made, or for their ways of incorporating them in the daily railroad-building process.

Fleming's own "passionate amateurism" lasted well beyond the period of youthful enthusiasm. The transitional moment between natural and rational can be documented in the "science" of Sandford Fleming himself, from the far side of the time divide, before his involvement in theory. His mind knew the science, but his heart still fancied Jules Verne.

In 1872 he gave a talk, titled simply "The Earth," before the Mutual Improvement Society of Ottawa. It is a sober and responsible talk, reciting all the known facts of nineteenth-century geology, astronomy, and meteorology: distances, temperatures, compositions. He quotes the English mineralogist (and clergyman) Buckland: "Next to the study of the distant worlds which engages the contemplation of the astronomer, the largest and most sublime subject of physical inquiry which can occupy the mind of man and by far the most interesting from the personal concern we have in it, is the history of the formation and structure of the Planet on which we dwell." Fair and graceful, and nothing, until late in the speech, would seem especially out of place even today.

In the midst of citing Humboldt's calculation of the earth's interior temperatures, a molten core 160 times hotter than the melting point of iron, and observing that the only solid part of the planet was the crust upon which we floated, and that volcanoes were the vents of this "awful fire" and earthquakes were caused by this "eternal ebulition," he suddenly breaks off from rational science into "natural" speculation. The line between *a posteriori* and *a priori* thinking, between objectivity and fanciful exuberance, was fine indeed, and continually wavering:

> This theory although supported by Humboldt and many philosophers of the present day, is at best a frightful one, and seems to be altogether contrary to reason. The conception of the boundary line betwixt the fluid masses of the interior and the outer crust of the earth, is most difficult for our powers of comprehension. How long could a sheet of paper exist on a cauldron of boiling iron, and yet the comparison is a just one: for the solid crust bears about the same proportion to the whole earth as a sheet of thick pasteboard does to a globe six feet in diameter.
>
> Can we imagine 160 millions of cubic miles of liquid fire beneath our feet, and yet the surface of our Planet remains tranquil, the seas and oceans cool, and our homes and ourselves undisturbed. Surely the interior of our globe must be made up of more peaceful materials. . . . [He then floats the proposition that the core of the earth is composed mostly of

water, and finds "that theory more consistent and simple and more agreeable to the feelings of man."]

From the parallelism of the opposite shores of the Atlantic, it is highly probable that the continent at some distant period separated from the continent of Europe and Africa and floated gradually westward. Indeed it is not at all unlikely that the continent of which this province forms a part is still moving from our fatherland. The difference of longitude betwixt this country and Europe has never yet been accurately ascertained, at least the observations of Geographers and Astronomers vary at different times. . . . If we allow half that difference of longitude for error, we will find even at that slow rate, forty thousand years would be sufficient to waft America from the shores of the old world, to its present position.

The action of volcanoes and earthquakes seems at first sight to be against the water theory, but then operations can be accounted for through the chemical agency of water and galvanic currents. With regard to this George Fairhold says, "We cannot consider the awful phenomena of burning mountains as more than superficial pustules on the mere skin of the earth." It is now pretty generally understood and acknowledged that water is one of the most active agents in the production of volcanic fire; and that all the active volcanoes now known are situated near the sea coast, and rarely, or never, far in the interior depths of large continents. We have very great reason to conclude that the utmost depths of volcanic action are not greater than from one to five miles. Catpaxi, in South America, is perhaps of all volcanic mountains the most distant from the sea, and yet is only one hundred and forty miles from the shores of the Pacific. This volcano from time to time throws up not only great quantities of mud, but also innumerable fish. . . . According to Humboldt, most of the volcanoes of America throw out vast quantities of water and slime, at one of the volcanoes of Trinidad, a white sea shark was picked up in the act of being thrown out with the mud; sufficient proof of a subterraneous communication with the sea.

It is fortunate for Fleming's reputation that he confined himself to the social aspects of engineering and time theory, where his occasional *a priori* enthusiasms were less in conflict with the inflexible precision of hard science.

Mathematical inferences had permitted the discovery of an unseen planet, Neptune, by Professor J. C. Adams of the Cambridge observatory, in 1845, plotted from the "perturbations" in the observed orbit of the remotest visible planet, Uranus. Spectroscopists studied fossil light from distant stars, establishing their chemical makeup, and that of the sun whenever eclipses provided the opportunity to photograph flares and the solar corona. Spectroscopy, in fact, became the most glamorous of Victorian sciences, as reflected in Whitman's "Learn'd Astronomer," earning for itself a special place in Huxley's celebration of nineteenth-century advances. It was reason's most abstract extrapolation.

Solar eclipses became obligatory, if often far-flung, convention centers for the world's astronomical elite to set up their lenses and cameras and wait for the perfect shot. The longest and most perfect eclipse of the nineteenth century, which occurred in the vicinity of Madras, India, in August 1869, attracted the world's leading spectroscopists, like Jules-César Janssen of France and Lewis Rutherfurd of the United States. The rare event attracted directors of national observatories from all over Europe and Latin America. Their comfort was overseen by the head of the Indian Army, General Strachey (of the celebrated Anglo-Indian family), himself a notable naturalist. Fifteen years later, at the Prime Meridian Conference in Washington, Janssen, Rutherfurd, and Strachey were delegates of their respective nations. Sandford Fleming would be present as a satellite British delegate, as would the plotter of Neptune, Professor Adams of Cambridge, and many of the world's observatory directors.

The chance to observe the 1869 eclipse was not confined to south India, of course. It also lured another future conference delegate, Cleveland Abbe, late of Russia, now the director of the Cincinnati observatory. Short on funds, he pitched his camp in the wilds of the Dakota Territory where he trained curious Sioux to help with the observations. The Indians, in turn, carved two commemorative sandstones with the date, "August," and "1869," which Abbe kept in Washington at the weather office the rest of his life. Those stones might symbolize the moment when the natural and rational worlds were briefly in balance. They might also signal the moment when native American science, starting from a deep deficit, had closed the gap with British.

ANOTHER HINGE moment between natural and rational thought occurred in the early 1850s, when the telegraph came to outlying Scottish villages. Country folk appeared with their messages tightly rolled, imagining they'd be able to jam them, literally, through the copper wires. Fathers offered their "wee-est lads" as would-be telegraphers, speculating that they'd be selected on the basis of stature to slip more easily through the wires to deliver the messages. But how quick was their absorption of technology, how seamless the transition from natural to rational. Within the magical decade of the 1850s, those lads from the same hardscrabble Brigadoons were filling the new technical colleges, designing and running the steam engines of the world.

Victorian England's technical and economic supremacy is applicable only to a single decade, that of the 1850s, the decade of Prince Albert's scientific leadership and the Great Exhibition at the Crystal Palace, against a long stagnation and decline, with a growing sense of inadequacy, even of panic. Prince Albert himself had stated Britain's preeminence succinctly, at the time of the exhibition: "Science discovers laws of power, motion, and transformation; industry applies them to the raw material." Steam to railroads. Electricity to copper wire.

Everywhere a Briton looked in the 1850s (except perhaps in the Crimea, or the Black Hole of Calcutta) he had seen reflections of British glory. Science and Industry. Empire and Progress. Charles Kingsley, one of the most influential of Victorian progressives, had written of his emerging generation in 1851: "The various stereotyped systems which they have received by tradition are breaking up under them like ice in a thaw, [and] a thousand facts and notions, which they know not how to classify, are pouring in on them like a flood." Difficult scientific texts, like Darwin's *On the Origin of Species,* were immediate best-sellers, and scientists communicated social ideas for the educated common reader through scientific prose, without compromise. Progressive thinkers regularly shared their ideas in workingmen's halls.

It didn't last. Just twenty-four years apart, two leading British scientists, both presidents of the British Association for the Advancement of Science, pronounced upon the state of British science. The difference in tone is telling. Sir William Fairbairn, in 1861, celebrated "the present epoch" as "one of the most important in the history of the world." (In the same year, a distracted America was launching its Civil War, "Canada" was just an uneasy linkage of Ontario and Quebec, and Germany was an archipelago of princedoms.) "At no former period did science contribute so much to the uses of life and the wants of society," Fairbairn said, then went on to quote Sir Francis Bacon, the founder of British science, for whom the "legitimate goal [of science] is the endowment of human life with new inventions and riches." The words fairly burnish the image of a self-satisfied, even smug, British ascendancy. There is also a debatable point to raise: Are "inventions and riches" the goal of science?

In 1885 Sir Lyon Playfair, reflecting on his long service to the crown and on the number of commissions of inquiry he'd chaired, contrasted British methods with those of the Americans (so robustly recovered from their Civil War) in assessing so mundane a subject as the decline of commercial fisheries:

> We go out and interview fishermen, who by their very nature know very little. They know the waters where they fish, they know the methods they've always employed, they know the raw tonnage of fish they expect to extract from the same grounds they've been using over several generations. The Americans, on the other hand, know that the fishermen are the worst possible source of information. America goes to its great pool of public universities, recruits the best men from every field, and consults the ocean itself, reads water-temperature, reads oxygen levels, studies predators, measures the size and condition of harvested species.

Oxford and Cambridge had been disastrously slow to introduce science and engineering courses to their curricula. Civil-service exams for the Foreign Service through the 1880s weighed Greek and Latin results higher (six and eight hundred respectively) than scores in the natural sciences (five hundred for chemistry and three hundred for other physical sciences). The cult of the gentleman still held sway. In 1885 the twin prides of Britain's ancient university system were judged not even the equal of second-rate German institutions. Science writers and popular philosophers such as Herbert Spencer harped on Darwin's darker prophecies of the "devolution" of species, and the "degeneration" of nations. Questions were being raised about the fitness of modern man and modern society, and the failure of British institutions to meet the competitive challenge of younger and more vigorous nation-states, particularly Prussia and the United States.

Nature still exerted a deep influence on Victorians, but it was a call that had to be tempered with reason. In particular, nature was only to be studied, not to be worshiped. Most of us have sampled the vast, cautionary literature of moral unraveling, in which reason and refinement "revert" to the state of nature (usually preceded by the adjective "raw"). The agents of degeneration, the dark legacy of Darwin's hopeful evolution, were everywhere. Gypsies, shamans, medicine men, "half-breeds" and "octoroons," Hindus, Catholics, Muslims, the miscegenist, the scholar who identified too strongly with his subject and "went native," becoming . . . a monster, a Kurtz, a madman. In the tropics we must always dress for dinner, maintain a steely distance from the natives, avoid dangerous spices, and keep our linens starched. The natives are children—appealing, sometimes clever, often mischievous, eternally in need of models and a stern rebuke. A stiff peg in the evening, preferably with quinine, will see us through.

After 1860 the contributions of the amateur, the "rambling" naturalist, began to fade. Cultured dilettantism showed its inadequacy, especially in the face of a technically proficient Prussia and an ever-curious, expansive United States. A genial agnosticism took root across Victorian middle-class society. Religion was a social duty, a weekly reach toward the sublime—rather like Fleming's "Earth" speech—but otherwise discouraged in the Victorian hurly-burly. This is not to suggest, however, that religious attendance or its attendant social pieties disappeared. Scientists still proclaimed their faith. Fleming, a faithful Scottish Presbyterian, rewrote his faith's prayer book to accommodate worshipers in isolated Scottish settlements in the western provinces, often hundreds of miles from an organized church and ordained minister.

Outward observation, all those Sunday sermons at sea, or in jungle clearings, remained a significant part of Victorian life, as of course were missions to the heathens on every continent. Sermons were crafted documents, listened to and judged not on their emotional content, but on their intellectual and moral merit. In the seventy years of Fleming diaries, no more than half a dozen missed Sunday services are recorded, and always with remorse and a good excuse. The fervor of old-time religion, the emotional faith, was forced underground.

Victorians wanted to know about the earth, the plants and animals, the nature of the sun and stars, the atom, and the human body, and they saw all of nature as part of a unified system. They believed in a knowable universe and the unification of all knowledge. Thomas H. Huxley, in his 1887 essay, "The Progress of Science," a review of science during Victoria's first fifty years, predicted the imminent announcement of a unified theory that would combine not only light, gravity, and magnetism, but also biology, linguistics, and religion. Under Victoria, he wrote, science had discovered the three touchstones of nature: the molecular theory of matter, the conservation of energy, and evolution. Those three principles were sufficient to anticipate more wonders to come. He defended the approaches of induction and deduction, wild guesses and speculations—"anticipations of nature," he called them—in a wondrous adaptation of time all his own.

Huxley's expectation of an imminent explanation was almost met. Eighteen years after his essay, Einstein's Special Theory of Relativity (1905) did grow from specialized concerns to have profound effects on philosophy and the arts, just as Darwin's had had thirty-five years before. What Huxley did not foresee—and this divides the Victorian from the modern world, the natural from the rational—was the resistance to reason. The first of *The Fundamentals,* the source of all twentieth-century biblical literalism, had already been published in 1902. This was a series of pamphlets, which appeared under the imprimatur of the American Bible League, which attacked humanism, socialism, feminism, and evolution, and promoted a new view of biblical inerrancy.

It was not just pride in the sheer accumulation of scientific knowledge that concerned the Victorians; they also demanded that it have practical application and social benefit. After meticulously listing and crediting the great minds of England and the Continent who had brought so much fundamental scientific knowledge to mankind since the Renaissance, Huxley proceeds to list their failings:

> . . . but weaving and spinning were carried on with the old appliances; nobody could travel faster by sea or by land than at any previous time in the world's history, and King George could send a message from London to York no faster than King John might have done. Metals were worked from their ores by immemorial rule of thumb, and the centre of the iron trade in these islands was still among the oak forests of Sussex. The utmost skill of our mechanicians did not get beyond the production of a coarse watch.

The Victorian mantra is simple and ringing: *Be of practical use to society, serve mankind.* What good is theory if it does not improve the lives of common man? This is England's great departure from the Continental tradition of theory and pure research. It is also an expression of the choice that was facing the United States after the Civil War—adopt the British or German model of higher education?—and it is evident in the thinking of everyone associated with the standard-time movement. Research is fine, ran the popular interpretation; just don't lose touch with reality. "The later Victorians as a group were men and women of remarkable moral resolution," writes Richard Altick. "Living in a wasteland strewn with blasted articles of faith, they carried on, with spirit and confidence. What they lost in intellectual assurance and emotional comfort, they compensated for in sheer strength of will."

Around 1870 the British technical advantage of mid-century stalled. Part of it was the inevitable result of competition with

Germany, France, and the United States, but Britain blunted its own growth by failing to increase its investment in technical and scientific education. The emphasis on social utility, the historic scorn for theory and basic research, cost Britain dearly. Sir Lyon Playfair noted that a single German university, for example Strasbourg or Leipzig, received ten thousand pounds more in direct state aid than the half-dozen colleges and universities of Scotland and Ireland together. Holland, with only four million people and four universities—the same in both regards as Scotland—outspent Scotland nearly fivefold. France, in her determined rebuilding after the humiliation of the Franco-Prussian War, had asked herself a question that Britain would not raise for another half-century: Why could superior men not be found in France at the moment of her peril? One obvious answer was that in 1868, the Sorbonne had received the equivalent of £8,000 in state aid for academic use. By 1885 that amount had climbed to over £3 million. The implications are clear: Britain had retreated into its historic insularity. The rallying voice of Prince Albert had long been stilled. France and Germany had acted on Albert's assertion that science and research are the source of wealth, power, and progress.

OF COURSE Victorians were not modernists. They could not know that their own proudest inventions—sociology and psychology as obvious examples—would contribute to the undermining of Victorian confidence. What did those new "social" sciences prove but that man behaved irrationally, and that one of the vaunted Victorian articles of faith, "character," derived as much from secret drives, repressions, or unconscious influences as from sturdy moral and social exempla? After the standard-time issues had been settled, Fleming spent the next twenty years perfecting a world-circling undersea cable that would link England instantaneously with all her distant colonies. He believed (and how could he not?) that intimate and immediate contact would spawn greater knowledge, loyalty, and affection between the mother of parliaments and her satellites. The better we know our government, the deeper our respect. How could they have guessed that loyalty and affection were better guaranteed by the mystique of remoteness, that intimacy was the surest guarantor of mutual contempt?

Their children became avid appreciators of all that the Victorians had dismissed as natural, or primitive. Hermann Hesse, Pablo Picasso, E. M. Forster, Romain Rolland, D. H. Lawrence, Igor Stravinsky, painters, dancers, and composers, tried to restore the balance between nature and reason and to inject, or even inflate, the primitive elements in their work. But by then it was too late. The "real" Italy, Tahiti, Mexico, and India had been tamed, even Christianized. Their mythic, sexual, and Tantric "natural" identities had been driven, Forster-style, into caves, or into the hearts of various darknesses. However faintly, however ghostly those longitudinal and time-zone borders were drawn, north and south, colonizer and colonial were pinned to the same grid, stuck in the same time. The natural world had been banished, except through the occasional sexual encounter with the Greco-Roman or Indo-Aryan ideal, or sentimentalized, or Romanticized, as in the case of Lawrence of Arabia.

The application of dispassionate reason to unruly nature created modern science and the methodology that still sustains it. Sherlock Holmes's famous magnifying lens can stand for all the microscopes and telescopes, the patient accumulation of "trifles" and the inexhaustible permanence of clues in the natural world still waiting to be discovered—and fashioned into stunning, all-encompassing revelation. Those methods of inquiry had created the Industrial Age and the technological revolution, they built the communications networks, they asked the questions that led to the discovery of bacteria, radiation, spectroscopy, natural selection, that spawned the fields of earth science, philology, physics, chemistry, sociology, and psychology. Call him Holmes, or Darwin, or Freud, or Whitman—or in the context of this book, Abbe, Allen, Barnard, Dowd, Fleming, Herschel, Janssen, Strachey, Struve—any of that largely self-taught Victorian band and their European and North American equivalents who applied the abstract to the practical, and the practical to the abstract, induction and deduction, to come up with a new way of organizing time.

Standard time is the biggest gauge in the world. It converts celestial motion to civic time. And the most important fact about Sandford Fleming is that he realized it first and brought it to the attention of the rest of the world. That the world would finally opt for a simpler configuration in no way negates his message.

IN 1870s Ottawa, where the social tone was set by the British-appointed governors-general, Fleming ranked first among equals. Sandra Gwyn in her history of early Ottawa social life, *The Private Capital,* describes Fleming as *sui generis,* topping the A-list of every ambitious hostess. He was everything the Victorian gentleman should be. In the city, charming, sophisticated, commanding, at ease in all circles, political or diplomatic. Camped in the extremes of the Canadian summers and winters, leading survey crews over the swamps and mountains, he reveled in hardship.

And yet, two years after his return to Canada from his sabbatical in 1878, he was relieved of his CPR commission. For a period of about three years, from 1878 until he righted himself in 1881, he exhausted the tide of good luck that had floated him from relatively uneducated, penniless immigrant to the number-one civilian in Canadian society. In the dark year of 1880, Fleming was reduced to begging Parliament for a fair settlement, citing his double commission on the two railroads, the workload he had carried, his broken health. In all his letters, only that of February 9, 1880, after his fate had been decided, can be called self-pitying:

I indeed felt the weight of the responsibilities that were thrown upon me and I labored night and day in a manner which will never be known, some time after I began to work double times. I had the misfortunate in two consecutive years, 1872 and 1873, to meet with serious accidents. By the first I came near to terminating my life, by the second I was placed on crutches for 6 or 7 months. During the whole of these periods except when actually confined to bed I never ceased to carry on my work which I need not say was at times very arduous. As a consequence my general health suffered and I was forced to seek for some respite.

It earned him a severance package (as we'd say today) made up of what he might have made as Chief Engineer of the CPR, alone, minus the salary he'd already received for his years on the Intercolonial, times his eight years in service. In the Gilded Age, when the golden goose took the form of the iron horse, Fleming was cut loose for $29,800.

The railroad was reconfigured as a private corporation under the direction of the Chicago-born William Cornelius Van Horne. Thus it was an American who saved Canada from American designs. (And the most designing "American" of all was the Canadian-born James J. Hill, head of the Minnesota-based Great Northern, who recommended Van Horne in the first place.) Under Van Horne, the intractable choices between northern and southern routes through the Rocky Mountains were resolved. Van Horne was efficient, which is to say in that era, sufficiently ruthless and brilliant to proceed rapidly. He also benefited from the principal lesson Parliament had learned from Fleming: that an undertaking the size of the CPR had to be privatized and insulated from political oversight. Thus even Fleming's greatest failure provided useful instruction. Note, incidentally, that when a second cross-country route, that of the Canadian National Railway, was extended to the Pacific, it followed Fleming's survey-route through Yellowhead Pass.

It was at the beginning of this dismal period that his time paper was rejected in Dublin. And then, his good fortune returned. George Grant wrangled a largely honorary appointment for him as chancellor of a Scottish-Presbyterian institution, Queen's University, in Kingston, Ontario. The job required his appearance on campus for no more than five days a year. He proved himself an exceptional fund-raiser, especially through his friend and fellow Fifer, Andrew Carnegie. He accepted directorships from the CPR and the Hudson's Bay Company. The rest of his life, for the next thirty-five years, was devoted to travel, writing, and lectures.

Fleming's harshest critic is Canada's popular historian Pierre Berton, whose *National Dream: The Great Railway 1871–1881,* is the most readable account of the political and financial intrigue that swirled around the building of the railway. Of Fleming's failure he wrote:

When the Royal Commission finally made its report it came down very hard on the former engineer-in-chief but by then the construction of the railway was proceeding apace. Fleming went off to the International Geographical Congress in Venice to ride in gondolas and deliver a paper entitled "The Adoption of a Prime Meridian." Greater glories followed. His biography, when it was published, did not mention the petty jealousies, the bursts of temperament, the political jockeying, the caution, the waste and the near anarchy that were commonplace in the engineering offices of the public works department under his rule. He survived it all and strode into the history books without a scar. The story of his term as Engineer-in-Chief is tangled and confused, neither black nor white, since it involved neither villains nor saints but a hastily recruited group of very human and often brilliant men faced with superhuman problems, not the least of which was the spectre of the Unknown, and subjected to more than ordinary tensions including the insistent tug of their own ambitions.

I will return to Berton's charges a little later, but observe here only that Berton, at his narrowest, is correct in his charges. Fleming's term of office on the CPR was a failure, and part of the responsibility for failure surely rests with him.

Berton's evocation of the mysterious "Unknown" lends itself to many interpretations; certainly he is referring to a host of problems, some of which were technical in nature, others political, and still others international. In general, anything undertaken by Canada is in some way informed by, accelerated by, or deformed by the the gravitational force of the United States. Like most anxieties, it can fuel ambition, or an extraordinary effort to transcend it.

Fleming's response to the aggressive presence of the United States was to resist it through a vigilant eclecticism, and to take from America, when possible, its energy and self-confidence. A sympathetic but fearful outsider like Fleming, looking southward in the early 1870s from his surveying encampment on the prairies, wondered to his friend George Grant if a more humane way of development than the American model of wholesale slaughter of all inconvenient human and animal life could not be found (though he was not overly optimistic). He marveled as well at America's energy unfolding across the continent, and at Canada's apparent inability to harness the same enthusiasm.

I N 1876 Fleming began to interpose himself into the standard-time debate through papers delivered at the Canadian Institute and his membership in a number of American engineering societies. He was no out-of-touch professor from a women's seminary in upstate New York; he was the friend of the rich and powerful in Montreal, London, New York, and Toronto, yet he still needed instructions on how to behave with the American power structure. Accustomed as he was to "memorializing" the governor-general and gaining royal assent for nearly any enterprise, and coming from a hierarchical society where he enjoyed instant access and prestige, he had to be advised not to depend on official channels. Abbe was his American mentor, and he knew whereof he spoke:

> It is not easy for those high in scientific or official Government positions to impose upon the business world any radical reform; the practical man rejects it as theoretical and the private citizen rejects it as an official impertinence by Government with his personal rights. Any law passed by our Congress would be apt to fall to the ground unless it has the hearty support of the people, the lawyers, and the judges. I doubt whether it is best to spend much time trying to force the American nautical Almanac to adopt the time reform. Better to move through commercial, not political channels.

In the comparative anarchy that was the United States, the appearance of a government endorsement could have negative effects on popular support. It is a practical lesson that Fleming immediately applied. He became something of a demon spokesman at chambers of commerce, cultivating dozens of such organizations in Canada, the United States, and, eventually, South Africa, New Zealand, and Australia.

Let me return, in some detail, to his first paper, "Uniform Non-Local Time (Terrestrial Time)," which was delivered before the Canadian Institute in November 1876, just three months after his return from Ireland. Like a first novel, it is rife with autobiographical elements. There's the recent Bandoran experience, the autodidact's historical research, and the idealist's impossible schemes. It is the *ur*-manuscript from which he drew all subsequent (and far simpler) refinements, but it shows that the idea for world time, which he later called "cosmic time," was with him full-blown from the beginning. It shows that Fleming's proposals owe nothing to Dowd, Abbe, or Allen. Truly, "time was in the air," and in ways that neither Mr. Myers nor Mr. Allen, nor even Professor Dowd, suspected.

The paper was divided into five parts. First came "the difficulties arising from the present mode of reckoning time," in other words, a world with too many times is a world without time at all; the familiar descriptions of "natural" and "conventional" divisions of time came second; third, a history of systems of time-reckoning, ancient and modern. It is the fourth and fifth sections, called "The importance of having 'Uniform time' all over the world" and "The practicability of securing all the advantages of uniformity, while preserving existing local customs," that concern us here and open directly into a universe both familiar and alien. Call it Victorian modernism.

Time and space were the true identities: twenty-four hours for the clock, twenty-four meridians dividing the earth. Why the hypothetical chronometer? It functioned as an impersonal time god, a kind of hyper-time. Fleming described the new hours:

> They ought not to be considered hours in the ordinary sense, but simply twenty-fourth parts of the mean time occupied in the diurnal revolution of the earth. Hours, as we usually refer to them, have a distinct relation to noon or to midnight at some particular place on the earth's surface; while the time indicated by the Standard Chronometer would have no special relation to any particular locality or longitude: it would be common and equally related to all places; and the twenty-four sub-divisions of the day would be simply portions of abstract time.

From his first paper on time-reckoning, Fleming was trying to separate the physical reality of time from the socially and psychologically constructed reality of hours. The numbered hours

("Six P.M., is dinner ready?" "Six A.M., have you done your chores?") are difficult to separate from local considerations—that is, from local or natural time—unless we change their names and our habits of association. Calling five P.M. by its new name of seventeen hours would help, but it was not enough. By proposing the concept of "abstract time," the hypothetical regulator set in the middle of the earth (or in the clouds), he was attempting to liberate time from specific locations:

> The standard time-keeper is referred to the centre of the earth in order clearly to bring out the idea, that it is equally related to every point on the surface of the globe. The standard might be stationed anywhere, at Yokohama, at Cairo, at St Petersburg, at Greenwich or at Washington. Indeed, the proposed system if carried into force, would result in establishing many keepers of standard time, perhaps in every country, the electric telegraph affording the means of securing perfect synchronism all over the earth.

The evocation of the telegraph—exact, instantaneous, and man-made—rather than the sun as a regulator of time marks Fleming as the protomodernist he was. It also alerted the French, who were more advanced in the applications of telegraphy to time than anyone else, to the presence of a man they felt they could work with.

The invisible, imagined, omnipresent "time-keeper" bears a strong resemblance to other Victorian deistic constructions. It, or He, is the invisible regulator of human affairs, otherwise tucked away and invisible. In short, Fleming was proposing a single time for the entire world, to be called "terrestrial" time or "universal time," so that when one's timepiece read "G.05," all timepieces in the world would register the same G hour. (It would not be G everywhere, of course; but every location on earth would know, as we know today by consulting the front pages of our telephone books, what time it is, relative to our-selves, anywhere else in the world.) If a train were departing at L.15, it would signify that it was leaving, relative to G.05, in four hours and ten minutes (no J, remember), and it would not matter if G were a morning hour and L an afternoon, since A.M. and P.M. were notations of the past. You were located both where, and when. Time and place were identities.

(In the theory's next incarnation, in 1878, Fleming proposed a modification of his twenty-four-hour clock. The hours from midnight to noon would read one through twelve, as they do now, but the P.M. hours would be marked by letters, starting with the same time-zone letter where one happened to be residing. Greenwich was Z, so that much of eastern North America, five zones earlier than Greenwich, was designated as U. Again, Flem-ing's pedagogical purpose was to fuse time and longitude, elimi-nating the social aspect of time. By 1880 his next revision had abandoned an East Coast landfall, and the Greenwich meridian as well.)

"Every traveller having a good watch," he wrote, "would carry with him the precise time that he would find employed every-where. Post meridiem could never be mistaken for ante meri-diem. Railway and steamboat time-tables would be simplified, and rendered more intelligible, to the generality of mankind than many of them are now." M.05 means merely that it is five minutes past the M hour, which follows L and leads to N. It is not impor-tant to know what the letters correspond to in old-style num-bered hours, since communications and schedules would only appear in the new style. One schedule would fit all listings, in all localities. There would be a separate reckoning for local time, which he did not propose eliminating—he just wanted to elimi-nate local time for any calculation beyond a strictly local applica-tion.

How would it work, practically? Or, more precisely, how would it *look*?

If a gentleman living in Philadelphia were to send a telegram to a relative in London announcing the very moment of his child's birth, in ways that would be mutually comprehensible, it might read as follows:

My dear brother Basil: At U:22 today 17 January 1881 our son Algernon Augustus III was born. (signed) A. A. Smith, Jr.

By such means, Uncle Basil in London, receiving the cable at Z:50, turns his watch face and learns that his bonny nephew was born a mere twenty-eight minutes earlier (in "real" or "universal" time). But without the meridional calculation, the time of transmission from Philadelphia, 3:22 P.M., would have no readily translatable equivalent to London's 8:50 P.M. By 9:00 P.M., Uncle Basil might cable his congratulations back to Philadelphia, where it arrives at, say, 4:00 P.M., local time. But the familial dignity of the occasion, and the technological achievement, would be obscured—literally—by the mutually unintelligible numbers.

Wouldn't it be more reassuring, more conducive to human understanding, for Algernon to receive Basil's wire at U:00 and know that his loving brother had responded to the news at Z:00, only minutes after having received it? As Frederick Barnard, the President of Columbia University, had noted at the 1882 meeting of the Metrological Society:

Mr. Fleming had caused to be constructed a watch to illustrate the proposed system. On the dial the hours ran from one to twenty-four, and surrounding the dial was a movable ring bearing the letters representative of cosmic time. By bringing the letters corresponding to any of the twenty-four standard meridians to the hour *twenty-four,* the watch will show instantly both the local time of that meridian and also the cosmic time.

(It may seem a simple-enough technology, another Victorian gauge, but it would disclose at a glance—and a twist of the wrist—one's local time relative to the time at any place in the world. The results would replace the bulky maps and notations, but the impulse anticipates today's cell-phone technology, which seeks a wireless connectedness at the flick of a button, for many of the same human and commercial reasons.)

Fleming was now through four of his topics (I've focused mainly on the fourth), but the fifth was, potentially, even more unsettling. In fact, it touched on a central conflict that would not be resolved until the Prime Meridian Conference in 1884, and then only in rancor. Part five: How to extend the advantages of standard time at sea?

"Navigators are required to employ a standard time to enable them from day to day, when on long voyages, to compute their longitude," Fleming began. The trouble, of course, lay in the multiplicity of prime meridians. Each ship derived its time from the prime meridian of the national observatory in the country of its registry. There were eleven national prime meridians—Paris, Greenwich, Rio, St. Petersburg, Rome, Lisbon, Cádiz, Berlin, Tokyo, Copenhagen, and Stockholm—from which ships derived their bearings at sea. Ships from different nations passing at sea could not communicate the location of mutual dangers, since British (and American) ships drew their charts and astronomical observations (their ephemerides) from a Greenwich prime, which was unintelligible to ships of other nationalities. The terrestrial marker proposed by Fleming would eliminate diverse meridians in favor of a single one, and, as he had already indicated, it really didn't matter which meridian that might be. (All nations, however, had to agree to it.) Yokohama or Greenwich, it was all the same, so long as it was consistently applied by ships of all nations.

The problem with Fleming's time proposals, even after he'd simplified them and done away with the imaginary time monitor in the center of the earth, and the dual-track clock-face, always lay with finding a universally acceptable prime meridian. All meridians were equal, they all measured the same rotation of the earth that we call a day—but some meridians were culturally and commercially more equal than others. Left to a popular vote, Greenwich would doubtless win, since 90 percent of the world's shipping already employed it. Even the American railroads, as well as the country's military and commercial fleet, ran on Greenwich time. But popularity alone did not necessarily recommend Greenwich to Fleming; quite the contrary, in fact, since the British meridian, as we have seen, lacked the neutrality he considered essential to a truly universal solution.

What Fleming proposed would be considered revolutionary, even today. Of course there would be a twenty-four-hour clock. There would be two time-tracks: local (such as we have today), designated by numbers; and terrestrial (which would regulate all maritime and continental rail activity and all communications), designated by letters. This, too, if we think of the airlines' Zulu time, based on Universal Coordinated Time, we have adopted for technical use. And here, Fleming remains our contemporary. Clearly, we are moving toward a single, uniting time. (The Swatch company, Swiss watchmakers, have even proposed an Internet Time that is also universal, allowing users in various parts of the world to bypass time zones and rendezvous in the same "real" time.)

That is, many millions need a local time for setting their dental appointments or movie starts, but otherwise live within a computer-driven universal standard of time. Cosmic time in the 1870s was a giant scheme for the capturing of real time. Telegrams would be sent at, say, M.13 and arrive at T.22, and both the sender and receiver would know precisely when they had been received. A telegram sent from London to, say, Denver, would show an apparent lapse of seven hours, and a few minutes. Actually, the two times are occurring at the same "cosmic" instant. The actual local time, either of transmission or reception, was irrelevant. Cleveland Abbe would not have to translate those dozens of weather-data telegrams into Washington time.

In 1876 Fleming had yet to work out the role of Greenwich (if any) in his universal scheme; hence that hypothetical timekeeper in the center of the earth. Clearly, given its "popularity," the Greenwich meridian could not be ignored in any eventual calculation. The challenge was how to make use of the Greenwich meridian *without* involving England. An interesting intellectual puzzle, and in 1878, when Fleming returned in earnest to the question of standard time, he came up with a suitably elegant solution. When is Greenwich not Greenwich? The answer is simple: when it is the "anti-prime" of Greenwich—not the zero degree of Greenwich, but the one hundred and eightieth degree, the continuation of the Greenwich meridian on the far side of the earth. In other words, Greenwich is not Greenwich—though it keeps all the ephemerides of Greenwich—when the anti-prime (or the "nether arc") cuts through "the unpopulated Pacific and over the icy steppes of Siberia," affecting no one, more or less, and arousing no national susceptibilities. The Colonial Office was suitably impressed, the papers were translated and circulated to the world's astronomers, and Fleming was invited to deliver the paper to the annual meeting of the British Association for the Advancement of Science, meeting that year in Dublin—and waited and waited, unsummoned.

A veiled demand for intellectual restitution is inherent in all of his succeeding papers, as well as in his final CPR engineer's report of 1879, in which he announced the idea of a sub-Pacific cable linking London through Canada to Australia. Railroads were fast, but the cable was faster, and Fleming had caught the fever of instantaneous connection. Railroads, he noted, were composed of two distinct technologies, the rails and the telegraph. They followed one another, and in fact rails could not function adequately without the telegraph. But cables were infinitely faster and more adaptable. The moment had come, now that the rails were in sight of the ocean, to continue the cables under the Pacific, just as they had already crossed the Atlantic. Vancouver would be connected to Fiji and Australia, Australia with India and South Africa. A glance at any map confirmed the fact that the red patches on the earth, the British Empire, fairly begged for connection. Without abandoning standard time, he would now take up the final great scheme of his life, the laying of the trans-Pacific and worldwide, all-British cable.

His vision had always been one of one-world and instantaneous communication. The time zones were but a rough sketch of what he next planned to do. What good is time if it can't be put to work?

IN 1895 Sandford Fleming, then sixty-eight years old and on the brink of achieving the great success for which he would be knighted two years later, was visiting County Mayo, Ireland. He was in Britain to check on the funding and the progress of his trans-Pacific cable. Of a small encounter that followed, he wrote:

On my journey in a jaunting car from Newport to Blacksod Bay at a wayside post office I telegraphed to a friend in London and proceeded on our way. In about an hour a woman appeared at the door of another wayside office. She hailed our car, and enquiring for a person bearing my own name, she placed in my hands a reply from my friend in London. The message I sent about eight miles back had crossed Ireland, the Irish Channel, Wales and England. It found my friend in the great city of London, and a reply was received in little more than an hour after I despatched my message, and the whole cost to me was sixpence. It was a marvel to me. Geographically I was in a remote corner of a country where I was entirely unknown, and I discovered myself telegraphically with my friends in London. Ever since my visit to Blacksod

Bay I have had visions of the extension of the use of the electric telegraph and have regarded it as a heaven sent means of communication. I have asked myself the question can we bring the Dominion telegraphically as near England as Ireland and Scotland are today? Can we bring the whole worldwide British Empire telegraphically into one neighborhood?

Miraculous as it must have seemed at the time, it is about as far as the marriage between steam and electricity can be pushed. The coordinated efforts involved, brought into focus that day, are also indicative of the mechanical disadvantage of steam technology, then entering its unacknowledged decline. The diesel engine had been invented two years earlier, the telephone was already widely in use, and the compact power of electricity was all the rage, from incandescent lights to the phonograph, oscillating fans, and motion pictures. Mr. Marconi's wireless transmissions would leap the Channel just a year after the worldwide cable was up (or, more properly, down) and running. (The wireless would bridge the Atlantic within the decade.) Had Fleming been listening, he might have heard scurrying sounds, primitive mice under the dinosaurs' feet.

8

Riding the Rails

THERE IS NO aspect of human activity, from law and medicine to economics and aesthetics that was not permanently altered by the encounter with our golem and faithful servant, the steam locomotive. We are fond of saying that the railroad "tamed the West," that it civilized the world, but there is a rakish counternarrative. Railroads emboldened us. The distant whistle fed our dreams, our hunger, made us, by prevailing standards, wanton. Its power knew no limit, and the power was transferable, right up the spine of the Victorian passenger. The problems they encountered learning to tame their master/slave—the speed, their dependency, the ecological damage—are our problems, too.

When speed becomes a consideration in all everyday decisions, it can be as unsettling, socially and psychologically, as speed in any athletic competition. Civil engineers, true to Sandford Fleming's description—the straighteners and levelers—had obliterated many of the irregularities of nature. The newly invented motorcar broke the hundred-mile-an-hour barrier in competition by the turn of the century. It is the perception of movement on all fronts, like skittering pond life, that defined the last two-thirds of the nineteenth century, and it is the railroad that lends itself as the single most conspicuous symbol of the Industrial Age.

The railway journey (to echo the title of Wolfgang Schivelbusch's great social history of nineteenth-century railway travel) was a microcosm of all nineteenth-century philosophical and aesthetic debates. What is vision, what is reality, what social structures can withstand the assault of speed, what are the fixed polarities of time and space? Who are our friends? What degree of distance is advisable? If the brain were indeed equipped with innate categories of time and space, as European philosophers contended, then the railroad represented a fundamental challenge to reality. The inner clock of the European psyche became unbalanced. How much of that imbalance expressed itself as hysteria and neurasthenia in the nerve clinics of early psychiatry? Schivelbusch points out that so-called "railroad spine," a host of symptoms without a definable cause—anxiety, sleeplessness, headache, loss of appetite, a generalized dread—yielded more to psychiatry than to traditional medicine, boosting the prestige of a field of study that had been notably lagging in popular esteem.

Even social etiquette came up for revision. In England, the easy camaraderie and crude democracy of the stagecoach that had led to those raucous *Tom Jones* moments of shared food and drink, and even to learning the coachman's name, never carried over to the railroad compartment. The reason, initially, was fear, and for good reason. Accidents, derailments, and general discomfort were common in the early years. One rarely chats or shakes hands on a roller coaster.

And there is another reason, one that reaches by analogy into the standard-time debate. The early European rail compartment was essentially, even to its velvet plush, a traveling coffin. No way in, or out, except through the individual door leading to the platform, and no circulation through the carriage between compartments. No on-board facilities. Heating was imperfect. We might even extend the metaphor to a kind of *No Exit:* purgatorial entombment with six cigar-smoking gentlemen, some, doubtless, with food and drink, reading their newspapers. No introductions, no stated common backgrounds or purpose. The first generation of European railway compartments were a chamber of horrors, a mode of speeded-up transport, but not yet a traveling experience (or as we'd say today, not yet a "culture of transportation"). Trains and passengers had not interiorized enough of the outer world—its sights, sounds and patterns of mobility —to make rail travel anything other than a disorienting hell on wheels.

The early years of European passenger-coach design exposed the central flaw in "natural" thought. Simply put, "natural" thinking failed to comprehend that a change in speed could not be separated from systemic upheaval, psychic rebellion. Railway carriage design in Europe owed much to the preexisting model of the stagecoach, but stagecoach atavism in an era of steam power was like repressive rule in a time of rising liberation. Speed and confinement are incompatible, except perhaps on a roller coaster. By the time the fear of speed and power was overcome (and adjustments, like the communicating corridor, were installed), a European railway etiquette of noncommunication had already evolved. Hence, the avoidance of contact, the sparse sociability, and the need (literally) for face-obscuring reading material.

In 1848 (in the wake of standardization along the Great Western), a newspaper vendor at Euston Station, Mr. W. H. Smith, began renting books to Birmingham-bound passengers. For little over a penny, they could take a book on board from his well-stocked library, read it, and then return it to his stand at the Birmingham terminus. Thus was a literature born, an extremely standardized literature, the so-called penny-dreadful, which took its place beside inexpensive editions of popular novels of the day. In France, Louis Hachette did the same, but on a grander scale, and soon British publishers were issuing special series of "literature for the rail."

Just as W. H. Smith adopted standardization for his lending library as soon as a common standard of time was struck throughout England, so did standardization encourage a unitary vision of culture, a shared cosmopolitanism across Britain and the continent. As early as 1838, ten years into the rail revolution, magazines were discussing the shrinking of time and space, and that the "national hearth" (meaning London) was now two-thirds closer than it had been to any of its citizens. Rails sped passengers along five times faster than the fastest stagecoach, which is a prescription for social hysteria, or another way of saying that perceived distances shrank by a factor of five. Schivelbusch quotes Heinrich Heine's almost rhapsodic 1843 embrace of the railroad's potential:

> What changes must now occur, in our way of looking at things, in our notions! Even the elementary concepts of time and space have begun to vacillate. Space is killed by the railways, and we are left with time alone. . . . Now you can travel to Orléans in four and a half hours, and it takes no longer to get to Rouen. Just imagine what will happen when the lines to Belgium and Germany are completed and connected up with their railways! I feel as if the mountains and forests of all countries were advancing on Paris. Even now, I can smell the German linden trees; the North Sea's breakers are rolling against my door.

And for every Heine, there was a John Ruskin who remembered the stagecoach and detested every minute of rail travel. In the *Quarterly Review* he railed (so to speak) against "the loathsomest form of deviltry now extant, animated and deliberate earthquakes, destructive of all nice social habits or possible natural beauty."

Both sides of the debate could easily project a time in the near future, for better or worse, when the entire nation would be one continuous city. Railroad travel rearranged social interaction, it stole the passenger's presumed autonomy. It forced upon him new ways of looking out a window (so as not to develop motion sickness). Instruction manuals taught him how to behave, how to safeguard his valuables, when not to initiate conversations. All new modes of travel and communication are briefly infantilizing, and railroads were no exception.

Much like today's generation, split between the computer-literate and unskilled labor, attitudes toward the railroad defined classes and generations. Flaubert was famously bored to distraction by railways, and Ruskin felt himself mishandled like a package, not served as a client. Others, however, took to the trains and welcomed the breakdown of barriers, even of landscape and all that was pictorial in nature, as a confirmation of the world as a unified, mental, symbolic reality. In 1848, in *Dombey and Son*, Charles Dickens, no early admirer of railways and their relentless gorging, like rampant brontosaurs on country lanes and urban neighborhoods, nevertheless honored their power, even their redemptive ability, in one extraordinary chapter, "Mr. Dombey Goes Upon a Journey." Dombey enters the train a defeated, near-suicidal wreck of a man. He alights, buoyed in spirit, freshly confident. Chapter Twenty is fiction of a high social-psychological acuity. Mr. Dombey, a wealthy industrialist, while preparing to board his train is accosted by Mr. Toodle, a "coal-raker," whose wife had been in service to the Dombeys. Mr. Dombey assumes he's about to be touched up by Toodle for a handout.

> "Your wife wants money, I suppose," said Mr. Dombey, putting his hand in his pocket, and speaking (but then he always did) haughtily. "No thank'ee, Sir," returned Toodle, "I can't say she does. *I* don't." Mr. Dombey was stopped short now in

his turn: and awkwardly: with his hand in his pocket. "No, Sir," said Toodle, turning his oilskin cap round and round; "we're a doin' pretty well, Sir; we haven't no cause to complain in the worldly way, Sir. We've had four more since then, Sir, we rubs on."

In other words, as Thomas Arnold had noted, "feudality is gone for ever."

In sociological terms, the class system was changing, and the mighty railroad was the cause. In psychological terms, Mr. Dombey enters his train compartment depressed:

He found no pleasure or relief in the journey. Tortured by these thoughts he carried monotony with him, through the rushing landscape, and hurried headlong, not through a rich and varied country, but a wilderness of blighted plans and gnawing jealousies. The very speed at which the train was whirled along, mocked the swift course of the young life that had been borne away so steadily and so inexorably to its fore-doomed end. The power that forced itself upon its iron way— its own—defiant of all paths and roads, piercing through the heart of every obstacle, and dragging living creatures of all classes, ages, and degrees behind it, was a type of the triumphant monster, Death.

Later in the journey, after his dark night of the soul, he comes to a tentative conclusion:

As Mr. Dombey looks out of his carriage window, it is never in his thoughts that the monster who has brought him there has let the light of day in on these things not made or caused them. It was the journey's fitting end, and might have been the end of everything; it was so ruinous and dreary.

All of which—the night aboard the train, absorbing his own kind of energy-transfer—permits him to rise the next morning "like a giant refreshed," and to conduct himself at breakfast "like a giant refreshing."

By the end of *Dombey*'s nine hundred pages, we feel that the author has likewise made his peace with the sheer brute majesty of the iron rails. Locomotion is the life force, *and* it is death; it is fate itself. Dickens might have wanted to pose as a Romantic poet or a Tory lord of a manor, and he would have loved to condemn all that destroyed nature and history, but he's Dickens; he cannot. He realized railways would bring life where previously there had been only despair and darkness. And just as surely, they will kill—famously, Anna Karenina and the eponymous hero of Willa Cather's "Paul's Case"—anyone who just can't adjust.

Railroads rewrote the law. Was the rail a "road" like a turnpike or canal, open to public use upon the paying of an appropriate fee? No, it wasn't. The British Parliament decided very early that the "rail-way" was something new and different. The right of way as well as the rolling stock could be deemed private property. Could they affect our health? The litany of physical complaints reads today like post-traumatic shock, and that exactly is what railroad travel was in the mid-nineteenth century, an uprooting of everything familiar, every notion of the tolerable, every received notion of time and space.

Eyes attuned to the pace and intimate perspective of the stagecoach were cautioned not to scrutinize roadside attractions but to focus on distant objects, the tallest tree, the church steeple or ruined castle, in order to avoid nausea, or worse, a disorientation that could lead to madness. Passengers had to develop "panoramic vision" to compensate for the disorienting fragmentation and treacherous glimpses of a blurred foreground. Inhabited landscapes, viewed from the moving train, dissolved into a series of blurs, mere impressions, shadows in doorways, distant shapes bent in fields, two-dimensional glimpses instead of long, per-

spectival approaches. This, as we shall see, fueled new aware-
nesses, new expectancies.

A WHOLE way of life had passed, as Victorian diarists were
fond of noting, a slower time of luxury and sociability, a chance
to know the intimate landscape from the stagecoach window,
when they swayed to the natural undulation of the horse. Obliterating inventions—the railroad over the stagecoach, the automo-
bile over the railroad, the typewriter over elegant handwriting,
the computer over the typewriter—often unleash such sentimental reveries. Who wouldn't prefer a train, a sailboat, a bicycle,
or cross-country skis to some jet-powered or motorized contrivance? Or a distinctive handwritten note from a fountain pen,
to some standardized, ill-composed e-mail?

We want more speed but we resent, or at least lament, the
elimination of the slower and, arguably, finer, more graceful experiences they replace. Railroad buffs, vintage-auto owners,
beer- and wine-makers, fly-tying anglers, gardeners—*they* are the
"temporal millionaires," who can afford to spend conspicuous
amounts of time indulging their fancies, living partially in the
past.

The rest of us buy upgrades on planes or trains in hopes of
restoring a touch of glamour, a bit of slower time, to an otherwise uncomfortable experience. For a stiff supplemental charge
we can imbibe again a whiff of Orient Express or *Titanic* luxury,
a bit of celebrity status the way Hollywood stars used to wave
coming down the airplane stairways as though they were political leaders. We can share a drink in the club car with Cary Grant
and Eva Marie Saint, or conduct black-market business like
Joseph Cotten, face-to-face, cigarette-to-cigarette, with Harry
Lime in the cold, cramped space between the railroad cars in
gritty postwar Vienna. Is that Bogart and Bergman on the tar-
mac, Claude Rains watching with a smirk? It wasn't so long ago
that all the drama, romance, and comedy of America was tied up

in trains, from Preston Sturges to Billy Wilder, all sung to the tune of "Chattanooga Choo Choo" and "On the Atchison, Topeka and Santa Fe." And the reason for that, I suspect, was because automobiles and airplanes had already "framed" trains, made them a site of memory and nostalgia, no longer a principal mode of conveyance.

STEAM TECHNOLOGY was about more than speed, power, and punctuality. Steam transformed more than the landscape. Steam was hot, loud, smoky, smelly, and dangerous, but there was also something intuitive about its working, and its direct successor, the internal combustion engine. One can imagine the 1850s version of the 1950s teenage grease-monkey, working on a steam engine, polishing, oiling, improving its efficiency. The leap from a James Watt to a Gottlieb Daimler or a Henry Ford is not unimaginable.

Steam was sophisticated, but apprenticeable. Unlike electricity, it was visible, a celebration of practice over theory. With steam, mountains could be bored and harbors dredged. Rivers were crossed, ships' designs turned from wood and sail to steel and iron, hold capacities and passenger cabins expanded a hundredfold, with a need to fill their holds with thousands of tons of coal for ocean passage. Meticulous planning and astronomical start-up costs entered the calculation for any new enterprise, and London and Continental banks oversaw bond issues for projects that might have seemed fanciful only a generation earlier: undersea cables, new shipyards, new steel mills, new mining equipment; transcontinental railroads spanning Canada, South Africa, America, India; telegraphs down the African coast. The new technology ran on coal, on more coal than traditional methods could ever extract. Workers had to be at least semiskilled just to handle the demands of the new technology, and, in the end, many grew sufficiently confident to challenge ancient wisdom, or to suggest shortcuts to greater efficiency. Quite a few, like the men who created time, learned enough on the job to become engineers themselves.

The traditional country blacksmith by his forge, the carter, the carpenter, required skills, but no skill that could not be passed on by way of apprenticeship. Imprecision was part of their rustic charm, and no one was seriously affected by minor imperfections. Until the 1830s, as Thomas Huxley and others pointed out, technological improvement and the velocity of events, the perceived speed of living, had lain dormant since the Renaissance. The economy was land, textiles, and agriculture. Imports and exports were triangle trades, carried on mostly by means of sailing ships, within the empire.

The technology of steam created the need for civil engineers to supervise the construction of tunnels, the laying of track, the deepening of harbors, and the analysis of subsoils for ever-heavier bridges—every practical application of engineering skills and applied physical and natural science. It needed mechanical engineers to set the minute tolerances and to oversee the manufacture of turbines and machine tools and to establish standard weights and measurements, and mining engineers to improve coal extraction, and metallurgical engineers to oversee new modes of steelmaking. Steam was an unforgiving power source: any weakness, any miscalculation, any misreading of the dozen or more temperature and pressure gauges could result in disaster. Stronger metals were needed, metals forged at higher temperatures, from purer ores, with experimental additives. The need for engineers directly benefited the middle classes and the great urban universities in London, Glasgow, and Manchester, not the aristocracy, whose sons headed for the prestigious universities and the more traditional professions.

In little more than a generation, England had gone from an agricultural and forest-based economy, unchanged since King John's time, as Huxley put it, to a rail-dependent, coal-based dynamo that was, by anyone's measure, demonic in its scale, noise, power, speed, and filth. Mountain ranges of slag ringed the new industrial centers. The long-range future of steam, however, was imperiled by its own inefficiency. The technology was condemned to gargantuism. The protective shielding and insulation required of a steam turbine, along with the difficulty of maintaining the proper pressure, made smaller steam engines uneconomical. Steam technology, therefore, developed noisily, sootily, even lustily. In the long run, the dimensions of steam applications were limited only by the availability of its energy source, and England, like France, Germany, and the United States, was endlessly blessed with coal.

The size and power of steam encouraged a swagger, a certain Gilded Age social and economic flamboyance, a cigars-and-brandy, godlike, frontier-pushing presumption of entitlement. The image of the glittering salons of the ever-larger steamliners and riverboats, of the upholstered saloon cars of the railroad elite, and the danger that lurked from sparks and boiler explosions—all fed a mounting psychic rebellion against restraint. The Gilded Age's desire to see more, travel farther, and go faster, in luxury and in freedom, with aggressive displays of tasteless consumerism, has to be set against the measured restraint of a Sherlock Holmes, or the staid images we've preserved of Victorian decorum. Both, of course, are accurate. The accumulation of new wealth—the emergence of the bourgeois model of Victorian affluence, so much a feature of literature on the Continent, as well as in England and America—is matched by the hazard of new fortunes, and the speculative losses that wiped out securities, and lives, as often as it created them. New wealth—dirty money, literally and figuratively—was won in high-stakes, high-risk operations in dangerous places. Bribes had to be paid, junk bonds floated, buffalo herds and recalcitrant Indians removed from the right-of-way by any means available. Consolidation and monopolization, starving out the competition, forcing mergers, calling in political debts, fighting the unions (most infamously by

George Pullman himself): it was a glorious time to be a buccaneer capitalist.

Towns became cities overnight. The rails brought immigrants, civilization, and new wealth. Growth, in the New World, whatever dislocation it caused, was viewed as superior to the emptiness it replaced, an attitude not unknown to this day. Demographics tell a tale, but so does myth. The myth of westward expansion had taken hold, manifest destiny was the call, the empty continent demanded settlers, and nothing would stand in its way.

B U T   I N England, where village culture had been established for centuries, and where the unspoiled country in some Arthurian, Shakespearean or Wordsworthian, leech-gathering, Constable manner defined the national character, rural survival required two conflicting conditions. Villages had to be "up to date," that is to say connected by rail and wire, *and* they had to persist unchanged in their nature. Could traditional country society, like "natural" thought in any of its manifestations, survive the industrial challenge?

The answer is clearly no. We've already cited Thomas Arnold's intuition, "feudality is gone for ever," from his first sighting of the railway in Rugby. D. H. Lawrence opened *The Rainbow,* which he was writing in the years 1912–14, with an evocation of the prerailway era in English country life that could have applied to Queen Elizabeth's, or even King John's time: "But the women looked out from the heated, blind intercourse of farm-life, to the spoken world beyond. They were aware of the lips and mind of the world speaking and giving utterance, they heard the sound in the distance, and they strained to listen." The contact with industrialism permitted the link between the "unspoken" and the world of "utterance." The city carried speech and mind, and spread contagions, such as land speculation, against which villagers were as vulnerable as Samoans before the common cold.

By removing horses from the power equation, steam began the long erasure of stylized nature itself from the countryside. Was a landscape still a landscape if it contained tunnels, bridges, a belching coal-fired locomotive, a railbed flanked by telegraph poles, and a string of passenger or freight cars, instead of a horse and wagon and the shell of a ruined castle? Was a village still a village if, gradually, it molded itself around the railway station and not the church or marketplace? What if, for survival, it learned to serve the needs and tastes of a few urban visitors, with inns and a semblance of city fashions? Could country life sustain itself if the young people were free to leave? And could an Englishman still feel English, still take for granted—as all great English authors had—a mystical connection between the national soul and stone-rimmed swards forever green?

Dickens, an *arriviste* country squire, came reluctantly to an answer. He viewed the scarring of the countryside and the effects of slash-and-burn railway construction in urban neighborhoods with a certain Tory horror. But he was Dickens. The squire might fret, but the writer was larger than his prejudices. Here was power as irresistible as flood or fire. If the energy released by technology could be internalized, and if the wealth created returned to society, poverty and dependency would end, and human beings would become supermen. His is the dominant voice of Britain's ascendant decade, the 1850s, when young Victoria ruled with the sophisticated Albert at her side, and Darwin launched the revolution in thought that continues to this day. The transformations might be ugly, the uprootings painful, the chaos at times unbearable, and there would be thousands of victims, but . . .

All that was stagnant in the countryside could be quickened, and all that was brutal and exploitative in the city could be overturned by new wealth, new visions, new possibilities. Lawrence saw it that way, as did Hardy. Power, nothing symbolic or mysterious about it, awesome, godlike power to move mountains, dig canals, alter coastlines, send goods to the best-paying markets, receive information instantaneously—all of this had happened suddenly, and no king, no priest, no self-appointed rural or urban potentate could contain it. Here was the magic link between wealth and progress that had proved so elusive, because technology, industrialism, steam, and sexuality (*and time! I would add*), were at base democratic achievements, not aristocratic playthings. Dickens, in time, applauded the change.

Railroads in North America did not disrupt the landscape, nor did they have to compete with earlier, and presumably finer, standards of civilization, or confront established patterns of feudality. There was no recognized history of place before railroads created it. Rather than challenge a prevailing orthodoxy, as they did in Europe, railroads were seen as spreading culture and values; they were, in fact, the free expression of that very culture. American travel had been defined before rails by river travel, and had accepted, as normal features of travel, the meandering pace of the river and the relative luxury of the river steamer. The European notion of travel was brief, direct, uncomfortable; the American, lengthy, leisurely, luxurious. Grand-trunk rail service of 1870 America is comparable to interstate travel by Winnebago today.

THE ARCHAIC, if charming, clutter of ancient *quartiers*, a series of medieval hamlets making up the metropolis called Paris, did not escape redefinition by the railroad model. Baron Haussmann, in his redesign of Paris (what David Harvey termed its "creative destruction"), used his once-in-a-millennium opportunity to impose a bold grid of straight, broad, railroad-style boulevards through the nooks and crannies of medieval Paris. Proud France had been humiliated by technocratic, militaristic Germany, and part of the blame for the scale of the defeat was thought to rest with the country's reverential embrace of its own self-regard. Europe's first revolutionary society had grown complacent. The *grands-boulevards* were part of a cultural makeover that was to have profound effects on French contributions to art and science, and would result in a dramatic, almost pugnacious, spillover at the Prime Meridian Conference in 1884. The division of opinion on the results of the redevelopment of the city neatly correlate with received attitudes toward railroads in general. You can have speed and efficiency, or you can have charm and tradition, but you can't have both. Haussmann at least attempted the delicate fusion of impossible demands.

The invention of standard time stands out as a defining act of social coherence, provoked by the spread of an invention that was slowly taking over our lives. Like the Chinese emperors of old, railroads bullied us, reigned over us, set their own clocks and calendars, created a handful of potentates, and left permanent clutter for others to clean up. Then we rebelled. The railroad was too important to eliminate, but we *would* put it on a strict schedule. We told it, literally, what time it was.

9

The Aesthetics of Time

The introduction of World Standard Time created greater uniformity of shared public time and in so doing triggered theorizing about a multiplicity of private times that may vary from moment to moment in the individual, from one individual to another according to personality, and among different groups as a function of social organization.

   —STEPHEN KERN, *The Structure of Time and Space, 1880–1918*

All language is of a successive nature; it does not lend itself to a reasoning of the eternal, the intemporal.

   —JORGE LUIS BORGES, "A New Refutation of Time"

CALLING A WORK of art timeless—for example, Nefertiti's head, Kafka's stories, Vermeer's interiors, Conrad's "Heart of Darkness," the stories of Borges—or remarking on its eternal quality, is high praise indeed. To declare a work dated is to draw attention to inherent elements that do not translate temporally, that limit its appeal or even turn it ludicrous to readers or viewers of a different era. Time usually renders such works melodramatic, sentimental, or simply irrelevant. However, there are those works that capture the spirit of their times in such a way that we praise their datedness and the dead-on accuracy of their observations. We might cite *The Great Gatsby*, or *The Sun Also Rises*, Thomas Mann's "Disorder and Early Sorrow," the works of Willa Cather, *Huckleberry Finn, The Secret Agent*, and, of course, Conan Doyle's Sherlock Holmes stories. There may even be a fourth kind of temporality in the arts, neither poignant nor timeless. It involves the atomizing of time itself, an experimentalism that actually endures, a *Ulysses*, a *To the Lighthouse* or *The Sound and the Fury*, the works of Proust and Gertrude Stein.

At the top of the stairs, and dominating the first of the French impressionist rooms at the Art Institute of Chicago, stands such a work, Gustave Caillebotte's immense (seven feet by nine) masterwork called, in English, *Paris Street, Rainy Day*. The "street" is something of a misnomer since it has been replaced by a bay of shiny cobblestones called Place d'Europe, the broad, sterile intersection of rue de Moscou and rue de Turin opened up by Baron Haussmann's leveling of an entire quarter. The canvas was finished in 1877, following several months of photographic preparation; the artist might have been working on it at about the same time Fleming was missing his train a few hundred miles away in Ireland. If the preexisting picturesque streets and quaint little corners of Paris had not been destroyed, we would not be talking of this particular painting.

Caillebotte's style has been called "urban impressionist" for its bleached palette, as compared to that of his better-known contemporaries. Along the right edge of the canvas a decorous gentleman and lady stroll under a shared umbrella, their faces obscured in shadow. The sky is ashy-lemon, a sulfurous brew of clouds and coal-fed pollution. Slightly behind them and to the left, half a dozen figures dash across the wet, reflective cobblestones. Under scaffolding, Second Empire buildings of the new Paris line the receding streets.

Only twenty-nine in 1877, and trained as an engineer, the bourgeois bachelor Caillebotte never had to sell his work in order to survive—indeed, he used his wealth to buy the unsellable canvases of his impressionist friends. A slight taint of dilettantism clings to his reputation: too rich, too sociable, too willing to sacrifice his own career in order to write notes and publicity for the cause of impressionism. He was not prolific; *Paris Street, Rainy Day* is one of only two or three Caillebottes that have made it into the canon of nineteenth-century French painting. In the judgment of Kirk Varnedoe he was neither the draftsman nor the colorist that his more celebrated contemporaries were; but for Peter Gay he was a painter of "distinctive, long underrated talents." In particular, Gay finds in Caillebotte's Parisian street scenes and bridges "breathtaking perspectives."

Paris Street, Rainy Day is not an ordinary street scene. It is something revolutionary, a sober portal in time standing outside Chicago's rooms of bright, impressionist light. One can infer many quasinarratives from that rainy Paris scene: the featured couple's ennui, their apparent lack of sexual tension, the stark soullessness—in the name of efficiency—of Haussmann's rational modernity. We can almost supply a caption to the scene, or devise an anterior story line. (*"Did you remember to turn off the gas?" "Where shall we dine?" "Do I dare . . . ?"*) *Paris Street, Rainy Day* foreshadows contemporary fascination with the not quite realistic world of Edward Hopper's paintings, or some of the apparently candid photos of Cartier-Bresson. It seems analogous to a great many literary works over the next half-century, those of Henry James, for instance, for whom objective reality practically disappears under the greater energy of subjective analysis. Or the reflective faceting of Virginia Woolf, in which the apparent coherence of a scene breaks down upon closer analysis into noncommunicating shards of multiangled perspective. They could have stepped from an early monologue of T. S. Eliot.

To emphasize a sense of urban alienation, or perhaps his own loneliness, Caillebotte had grafted technology to his palette, taking photographs of pedestrians at random intersections and gridding them onto his canvas. The result is thus constructed or compiled, not organic. The distorted planes are not the result of faulty draftsmanship but are deliberate—Peter Gay's "breathtaking" is not at all an exaggeration. The half-dozen figures crossing this cold, wet urban space are *truly* not related. Each figure seems like a monad in space clinging to a separate plane. The bored couple stroll purposelessly in the rain. No figure is posed. The presence of a painter is not even implied.

In the old Dutch interiors, particularly Vermeer's, so precise is the compositional geometry that critics have been able to triangulate from the two-dimensional plane of the rendered figures to an implied third dimension—that of the artist himself. They can speculate on his distance from the subject, his height, even whether he had been standing or seated as he painted. Perfect perspective derives from a point of origin as well as a vanishing-point, all coming together at the centrally placed focus (in the case of Vermeer, the pouring of the jug of milk, the sheet of paper, the unfurled map), and the play of shadows from an exquisitely captured light source, a half-opened door, or narrow overhead window. That comforting perspective is missing in Caillebotte.

Of all the fine arts, painting would seem the most immune, or to have the least relationship, to time. The standards up to about 1875—in subject and execution—had been set forth by the Salons and the Academy, reinforced by critics and museums. Subject matter was suggested from myth, antiquity, national history, and religion, or from exercises in anatomy, still lifes and figure studies, outdoor vistas, street scenes, always with demonstrated mastery of technique: shadow, folded cloth, clouds and water. Commissions derived from flattering likenesses and richly rendered interiors. Prizes were awarded on demonstrated mastery of composition and color—color, that is, true to "nature" as captured by the human eye.

Fundamental questions had been opened up by new sciences, or, to put it in terms of this book, imperfect welds had been loosened by new velocities. How is reality really composed, how is consciousness constructed? We can't trust our eyes (Muybridge and Marey showed us that), and if we can't trust what we've just seen, can we rely on our memories? How to explore the opened-up gap between perception and reality? Cubism, to take a later example that seems peculiarly susceptible to temporal analysis, replaced the old consecutivity and painterly narrative with simultaneity. The past, the present, even the future, happen at the same time on the same unmodulated plane of color. Breathtaking perspectives in all the arts, I would argue, are the result of the new consciousness of time.

The birth of modernism is a story well and frequently told: how Paris and Vienna and Berlin arrived at a new awareness, and how the intuition spread universally; how the structure of belief and notions of reality were revealed to be the unexamined residue of received opinion; how inherited forms were found to be arbitrary and outmoded (unreadable, unwatchable, unviewable, sentimental, and false). Artistic manifestos called for the disintegration of traditional form in all the arts. The cracks and fissures that Victorian culture had ignored, thinking them mere imperfections, that troublesome scurrying noise at our feet, could no longer be left out of the calculation. In fact, the imperfections became the whole story. Pursuing those cracks and fissures wherever they led opened up a virgin continent of uncertainty and the unconscious for exploration. Reality was, at best, an appearance, a reconstructed patchwork, an artifice.

The hierarchical world of painting imploded. Revolutionary conditions that had applied to the railroads and the sciences had arrived in the art world. France was once again a rising power with a confident bourgeoisie amply served by a coterie of taste—

its galleries, dealers, and exhibitions. Paris had been "modernized," as though a new time zone had sliced through the natural, medieval miasma of ancient streets.

CAILLEBOTTE'S COMPOSITION is chillingly impersonal, the figures anonymous. There is no anecdotal focus of interest. The viewer's eye (mine, at least) is thus drawn to—of all things—the strollers' feet. No pair of feet of the human figures is firmly planted on the pavement. If the heels are touching the cobblestones, the toes are up, or vice-versa. Each figure appears to be in motion. This obviously ungainly assault derives not from Salon standards but from the example of candid photography. The painter cannot "see," truly see, how people walk; the eye blends the movement into one continuous action. And that is precisely the point. *Paris Street, Rainy Day,* with its muted colors, is the quietest impressionist painting in Chicago, but also the most dynamic.

The impressionist revolution, we've been told, is all about light. Light means a self-lit subject, liberated from an external (or at least, identifiable) light source. It means overthrowing formalities—those of the posed subject, the play of light and shadow, and the illusion of perspective. Even when the impressionist subject was taken from nature, as so many were, it is a recomposed nature, nature stopped, stripped of academic signs of "naturalness," and then re-seen. The impressionist canvas is nonnarrative (as its casual or flatly factual title often indicates), antinarrative, amoral, it flares on the wall, commanding the eye even before we figure out what it is "about." It doesn't "mean" anything, it is not an "illustration of" or an "allusion to," there's nothing to "get" except the pleasure of performance. And because of that, its significance is profound.

Impressionists understood intuitively that energy—and it was energy that became the art of the age—was conveyed by radiated light, not by well-rendered form. Light is energy, as Einstein eventually showed, and to the impressionists, the Salon standards of form (like the Newtonian universe for Einstein) were the first impediment to the release of energy. Energy is released when light scatters form, not when form insists on its own perfection. It was a bold step for artists to take, but the only one in keeping with the times ("time was in the air"). It was alienating to the public at first, and to most critics, and it posed a challenge to received opinion and the finest standards of instruction that the Salon, quite properly by its lights, did not accept.

The success of their revolution today renders a perfectly realized Romantic masterpiece like Géricault's *Raft of the Medusa* dated, with its anatomically correct, tortured bodies on a storm-ravaged sea. The figures seem at best melodramatic, at worst cadaverous. Although the anecdote of the wreck is dynamic, the figures seem posed. The reason Caillebotte's painting is provocative at nearly any depth of interpretation is precisely *because* it has no anecdotal subject. It is "about" a Paris street in the rain; that is, it is about nothing anterior or paraphrasable. It may be incidentally about forms of urban bleakness and sexual alienation, but it embodies something far grander: the capturing of a single anonymous moment on the tapestry of time. Like Bloomsday in *Ulysses,* or June 2, 1910, in *The Sound and the Fury,* the day of Quentin Compson's suicide at Harvard, it is everything—and nothing. Any random day is shown to be, upon close examination, a microcosm of all human history.

On his death in 1894, Caillebotte willed his collection of sixty-seven canvases by Monet, Renoir, Degas, Cézanne, Sisley, and Manet to the Louvre, only half of which, because of Salon pressure, were grudgingly accepted. France's loss is the world's gain. Those impressionist rooms around Caillebotte's *Paris Street, Rainy Day* at the Art Institute of Chicago today suggest an expanded version of the walls in his own Paris studio, crammed with Salon rejects he'd bought from his friends a century ago.

In *The Bourgeois Experience: Victoria to Freud,* Peter Gay

reads a deeper complication than mere jealousy in Caillebotte's bequest and the Louvre's reluctance to accept it. It is yet another form of the rift between the received and the revolutionary, or, in terms already described, the natural and the rational:

> But the fundamental question of choice—whether to preserve the safe or make way for the risky—made for frictions that necessarily played themselves out in public disputes. And these disputes were, it seemed, growing less and less reparable. With the coming of modernism the breakup of a great if shaky bourgeois compromise was on the horizon, a compromise that had held culture together for decades.

Caillebotte is an attractive and humane figure, like Pissarro, Seurat, Monet, or Cézanne, or Henry James, Mark Twain, Edith Wharton, or Gertrude Stein, one of those apparently stable, confident figures we associate with the late-Victorian, Edwardian, and pre–World War I era who seems capable of living out Flaubert's maxim to write (or paint) like a demon, while living like a bourgeois.

And there are others, like Fleming, and Cleveland Abbe, and the Parisian spectroscopist Jules-César Janssen (who created the Paris Observatory in those same years), who are part of the same society, men and women of easy graces who were not reclusive or revolutionary but, in their way, bomb-throwers. Rebellion (or, as they saw it, reform) burned on a low flame within them, but they were nevertheless the loyal opposition to many received opinions. Cleveland Abbe worked tirelessly with the Freedmen's Bureau and with General Oliver Otis Howard (the founder of Howard University) for what was called "Negro uplift." The woman he married had spent the post–Civil War years as a volunteer teacher of Negro children in Mississippi. Abbe himself attended black churches during his years in Washington, and even brought Sandford Fleming with him when the latter was in town during the Prime Meridian Conference.

In their personal lives they seemed to be able to contain conflicting beliefs in faith and science, in authority and freedom, in charity and acquisition, and, in some cases, between probity and abandon. They trouble us today, with our automatic association of modernism with rebellion, exile, imprisonment, and alienation, and of Victorianism with bland hypocrisy and guilty repression. They were radical moderates.

THE STIRRING in the arts called modernism—"the breakup of a great if shaky bourgeois compromise," in Peter Gay's words—is really the implanting of time inside all artistic constructions. And it is time, in the form of velocities, that forced the breakup. The Peruvian novelist Mario Vargas Llosa claims Flaubert as the father of all modernists: "With Flaubert the novel appeared for the first time not only as a moral task or the creation of a story but also as a purely technical problem, the problem of the creation of a convincing language and the organization of time and also the problem of the function of the narrative in the novel." Raymond Williams made the same claim for Dickens. He did not *report* on social change (as George Eliot had), he found ways of embodying it in plot and character. In impressionist terms, they found ways of capturing an internal light source (time); not recording the illusion of passing time, but of entering it.

Embedding the narrator inside the novel as early as the 1850s, is at the very least protomodernist—but also temporalist. Flaubert had to construct a time machine for *Madame Bovary* and, very consciously, find ways through voice and language to place a story inside it. We are back again in that moment, identified by Henry Adams in 1844, when the world changed perceptibly, due to railroads, ships, and the telegraph; or perhaps it is the moment of Britain's standardization, noted by Dickens; or of Thoreau's retreat to Walden Pond. We are back to the generation of Melville's Bartleby, and the struggle to mold time, language, and character into a single coherent story. We are back to Fleming's "chronometer" buried in the middle of the earth.

The photographer Eadweard Muybridge with his horses, the motion-picture pioneer Étienne-Jules Marey with his blended vision, the physicist Ernst Mach, who caught the muzzle discharge of a bullet on film, the experimental colorist Georges Seurat, who demonstrated the eye's subservience to the brain in mixing color (his *Sunday on the Island of the Grande Jatte* hangs a few rooms away from Caillebotte's *Paris Street, Rainy Day* in Chicago). The impressionists and their successors all broke "natural" movement into stills, then recomposed the disjointed frames into the semblance of unity, or of motion.

Technical inventions such as Frederick Taylor's stopwatch and the high-speed camera demonstrated our inherent physical frailty, our unconscious reliance on habit, and our physiological capacity for self-deception. Those vulnerabilities foretold the intellectual enterprise of the new social sciences—sociology, anthropology, scientific management, political science, psychology. For the most part, the social sciences brought unwelcome news: we're less in control than we thought we were; less free, less virtuous, less enlightened. They also broke down the apparently natural continuity of observed human activity and interaction into discrete stills, stopped-time frames that would, ideally, yield up a clearer picture and deeper understanding of our unconscious motives. The ideal mode of analysis for sociologists and psychologists is the candid observation (Taylor hid his stopwatch inside hollowed-out books), the ability to stop time, to interrogate the "natural" with theories of the rational, like the skinsman in jazz.

Paris Street, Rainy Day would not convey its special tension if the feet of its subjects were fully in contact with the cobblestones. It would not *inhabit* its scene, à la Flaubert, if the various subjects had actually existed that day, just as the artist saw them—it would have been, rather, just another composition *about* a Paris street in the rain. In 1877, photography and impressionism came together in Caillebotte's studio. Light and time were captured and released. The Caillebotte composition is Flaubertian, not merely representational.

The need to manipulate time is central to every technical, intellectual, and artistic discovery of the twentieth century. Technologies seek to "save" time by speeding up connections or increasing efficiency and load capacity; artists look for ways of extending the moment, "saving time" in the sense of preserving it. Stopping time, extending the perceived present and challenging the "flow of time" (a notion as old as St. Augustine), defying the Salon palette and its disciplined command of highlight and shadow, is the aesthetic counterpart to the standard-time movement. Artists, writers, and scientists all learned the tricks of temporal manipulation. They absorbed the central lesson of the temporal revolution: time is not given by God, it is taken by man. Bell and the telephone, Otis and the elevator, Edison and the lightbulb, Pullman and his luxurious dining and sleeping cars, Caillebotte's breathtaking vistas and Seurat's adaptation of psychochromatic theory—all of their inventions and masterpieces manipulate time. Pursue any of them, *a priori* or *a posteriori*, and you challenge the limits of natural, local time. The time zones themselves are innovations of the same Industrial Age. Each hour punches in like a worker at a time clock. The earth revolves like a giant cog, its teeth biting into a larger cosmic wheel. The neat divisions between the zones have the effect of compartmentalizing time arbitrarily, as though we are capable of keeping only one hour at a time, a thousand miles, fifteen degrees of longitude, in our minds. What standardization did was reduce the number of time standards in the world downward from infinity to twenty-four. It expanded the bubble of identical time from twelve miles to a thousand. (The next step will be to reduce the Industrial Age's twenty-four down to one.)

Artists took the temporal revolution much further than the engineers intended. They reacted to rationality, especially to the smug rationality imposed by industry and capital. Standard time, once it was constructed by diplomats, astronomers, and engineers for the benefit of industry and management, had to be artistically, subjectively, demolished. True clarity and understanding could be achieved only in the deliberate restructuring of time and space. Liberation from nature and religious dogma is a fine goal, but slavery to a time clock is no great improvement. The dreadful progress of Victorian will and order could be opposed most effectively not on the streets, nor in the bars and bedrooms, but in the ateliers and garrets, by subverting the very tools of logic itself in the artistic derangement of form and language.

The touchstone literary confirmation was supplied by Joseph Conrad in *The Secret Agent* (based on an actual event of 1894), in which a band of anarchists set out to destroy the viability of British society, not by the bombing of Buckingham Palace, the Inns of Court, or the Houses of Parliament, but by blowing up the Greenwich Observatory. "The attack must have all the shocking senselessness of gratuitous blasphemy," explains Mr. Vladimir, their philosophical leader. In a mercantile society, a single, unified time, everywhere and indivisible, is the invisible but omnipotent God. It must be assassinated. Conrad's anarchists, though extreme in their planned violence, capture the modernist tone exactly: opposition at any cost to the established order. The fundamental order is temporal.

THE MODERNISTS' ways of rendering reality make our *fin-de-millennium* writing seem simplistic by comparison. James, Woolf, Conrad, Lawrence, Joyce, Stein, Pound, Eliot, Dreiser, Hemingway, Faulkner—just staying within the English-language tradition—pose difficulties we're still trying to unravel. Add the others—Proust, Mann, Kafka, Musil, Broch—and readers of serious early-twentieth-century fiction are confronted with an apparent epidemic of temporal obsession.

Time was in the air. New time standards had swirled around the childhoods of Mann and Proust, Woolf and Kafka, Einstein and Joyce, all of whom were born in the Decade of Time. Their works are all about time, but standardization addressed only the most superficial complaints of temporal instability. Standardization rationalized time for the industrial worker and the railroad passenger and the managerial elite, and made the same laws and the same time apply over wider and wider areas, but it did not eliminate temporal anxiety, especially in the various artistic undergrounds. In fact, it freed them to call attention to an unfinished temporal revolution. Standard time could not penetrate the subjective layers of memory and repression. Exposing the layers of repression and secret motivations became the new literary and theatrical agenda.

In 1885, while still working in Antwerp, Vincent van Gogh began decorating his studio with Japanese prints of "little women's figures in gardens." When van Gogh arrived in Paris two years later, he invested all he could in Japanese woodblock prints. Their appeal, he explained to his brother, lay in the unmodulated planes of color and exaggerated perspective. That same year, he organized two exhibitions of Japanese woodblock prints in a Montmartre café. Japanese style can be seen in the bold, unmodulated colors of the "Irises" series, which he began in Arles the following year and continued in the urgent last paintings before his 1890 suicide. "Unmodulated" means unshadowed and nonperspectival; color dominates form, urgency transcends pictorial illusion.

Just as Victorian rationality had led confident researchers into the depths of the irrational, so did temporal freedom lead artists to the distortion of all received notions of social and psychological reality. Painting took on the stark geometry, the chaos and the commercial glitter of the urban palette. The arts in general—painting, writing, music, dance—tried to get inside the pulse of the city. Entering the frame of the story, discreetly, like Flaubert, seemed tame indeed compared with the appropriations of *Ulysses*. Even the brightest impressionist paintings paled against the colors of van Gogh. Cubism in painting and its extensions into poetry and fiction collapsed the three-dimensional time-space continuum into two. "Not seen in nature," which might have been a charge of opprobrium in an earlier era, became the new sign of genius.

And the cities themselves were transformed, packed with newly liberated, newly empowered, freshly expectant immigrants. Their pasts wiped out, their futures were suddenly before them. The stories by now are familiar, how a window of tolerance beginning at the start of the century permitted Jews from eastern regions to find legal residence in Vienna and Berlin, how the generational leap from *shtetl* to university—from Talmudic to secular learning—unleashed an energy source that had lain dormant, and helped bring about the Continent's intellectual rebirth, and how others from southern and eastern Europe, Catholics and Jews, did the same for the open cities of the New World.

T H E   I D E A of the modern has undergone considerable revision in the past several decades and, on the authority of William Everdell's *The First Moderns,* can now be pushed back to the mid-1870s, and to fields far outside the arts. The dyed-in-the-wool Victorians, by comparison, those progressive thinkers of an earlier era, like Fleming, remained resolutely objective, impatient with subjectivity, suspicious of private emotions and unsupportive of arts that threatened to undermine confidence or grow morbidly introspective. In 1878, at a time when Fleming's London circle of contacts had expanded, and when his colonial timidity had sufficiently receded, the cultural figure he chose to visit was his aged fellow Scot and one-time Kirkcaldian, Thomas Carlyle (born 1795), a coeval of Keats. They spoke not of Carlyle's darker fulminations but of memories of Kirkcaldy, and of stirring events on the Canadian Pacific Railway. For engineers like Fleming, the objective world of undiscovered and unexplained nature, not the seething unconscious, still beckoned.

FAULKNER AND Hemingway were born a year apart at the close of the nineteenth century. Their names are inevitably linked, their achievements endlessly compared and contrasted. One wrote memorably of suicide; the other committed it. One captured America's celebrity fancy; the other rarely traveled and is associated with only one town, one state, in which he led a reclusive life. And later judgments on both have been harsh: on Hemingway for his macho posturing, on Faulkner for his resistant regionalism and the racist residue it included. Both were enamored of time, and took their fascination in opposite directions. In the case of America's two greatest writers of the twentieth century, the ancient gods of time looked forward to death, or backward to history.

In Our Time (not to push the title) is the most influential collection of stories in American writing. With *Dubliners* (and, to stretch the same point, *Winesburg, Ohio*) it defines one polarity of modernism. Fragments of time, place, and character—with none of the logical or linguistic unity of an earlier age—are narrated in the clearest and simplest of language. Hemingway's famous style denies continuity; it fragments time, sentence by sentence:

> He liked the girls that were walking along the other side of the street. He liked the look of them much better than the French girls or the German girls. But the world they were in was not the world he was in. He would like to have one of them. But it

was not worth it. They were such a nice pattern. He liked the pattern. It was exciting. But he would not go through all the talking. He did not want one badly enough. He liked to look at them all, though. It was not worth it. Not now when things were getting good again.

(from "Soldier's Home")

It was raining. The rain dripped from the palm trees. Water stood in pools on the gravel paths. The sea broke in a long line in the rain and slipped back down the beach to come up and break again in a long line in the rain. The motor cars were gone from the square by the war monument. Across the square in the doorway of the café a waiter stood looking out at the empty square.

(from "Cat in the Rain")

Only one comma intrudes in the two selections. Commas are like Band-Aids, protective adhesions permitting transition, and here none are needed. Hemingway's prose style of the early 1920s owes to his journalistic training and obvious opposition to Edwardian fussiness, but also to the internalization of death or, in my terms, the anxiety of time. (No wonder he admired Gary Cooper as an actor, the same shuddering delivery, courting slowness, emphasizing the discontinuity between lines.) Time is present in the stuttering progress of the sentences themselves, stopped-time frames of activity where self-consciousness intrudes to disrupt all continuity. His sentences, devoid of shadowing, are the literary equivalents of van Gogh's unmodulated planes, or of Caillebotte's strolling Parisians. They dazzle but do not illuminate, they reflect light but surrender none of their own. They retain their meaning, spitting out only the seeds. His characters, at least in the early stories, are devoid of a temporal element; they are (like the girls in "Soldier's Home") part of a pattern, but not part of this world.

Those simple, repetitive sentences with virtually no movement between them are the metronomic ticks of a life running out. Hemingway often stated that death was dealt a hand at a serious writer's table. His sentences claw for a grip, reaching for a future that's forever retreating.

STANDARD TIME, as Conrad implied, is a secularized religion. Time has a moral character; our conscience keeps it eternally present, it won't go away, won't bury itself in the past, where it belongs. When we think of time, of histories and cultures and lives that have preceded our own, when we hold their artifacts, or read their records, we are likely to be filled with a secularized form of awe that is akin to worship, what the religious feel in the presence of God. Like Simon Schama, when we think of time, we're brought back to our personal Pook's Hill, where "lucky Dan and Una got to chat with Viking warriors, Roman centurions, Norman knights, and then went home for tea." Or, if not for a cup of tea, to a shot of bourbon, to Yoknapatawpha, ruled by an unforgiving time lord.

The nineteenth century struck down God but didn't bury him, as my friend, R. W. Rexford, likes to say; it erected standard time in his place. Works of art that take time as their theme are sublimated works of religion, taking the Bible as their narrative frame. (How did we start so innocently and become what we are? How did this social, political, environmental mess come about, and how can we atone for it?) The past is never over, it's continually enacted, like the herds and priests on Keats's Grecian urn.

In Faulkner, modernism reached its American apotheosis, at least in terms of temporal derangement, as well as in overt time consciousness. The two faces of time are fused, the dead come to life, the past becomes again the present, the "backward-looking ghosts," as Quentin Compson calls them in *Absalom, Absalom!* die again for their sorry failure to redeem a precious drop of spilled, ancient blood. In *Light in August* Joe Christmas, being run to ground by a lynch mob, reflects on the sheer dumb majesty of being born *here,* and *now,* and not some other place in the tapestry of time. The past is alive, it is palpably present, because the present, the contemporary characters, have no moral force, no sustaining life.

The Sound and the Fury can be read temporally as a war between the blighted "natural" world of the "idiot" Benjy and the narcissistic, hyperrational world of his brother, the Harvard student, Quentin, against the ferocious mechanical reductionism of their brother Jason. Benjy lives in an eternal moment, undifferentiated by civil concerns. Quentin exists on the day of his suicide in a temporal straitjacket he's trying to escape (which, of course, he does), and Jason in a crude rationalism that he turns to pitiless power and profit. By ending the book on the character of Dilsey, Faulkner implies there's yet another way out of the three-cornered temporal tragedy, but it's the way of endurance, of "prevailing," the way of love and patience and forbearance, the exercise of mammy-spirit.

In Faulkner's ethos, rationality is a fatal disease, for which the only antidotes available to white Southerners are idiocy or cold, calculating brutality. Quentin wants to escape civil time, the reminder of his sister's marriage and his shame, and to enter eternity. He smashes his pocket watch, and avoids looking in the jeweler's window, which is filled with sample clocks. "Because Father said clocks slay time. He said time is dead as long as it is being clicked off by little wheels; only when the clock stops does time come to life."

What Faulkner describes is the Southern tragedy in temporal terms. There was a natural time and a natural order, the world of the forest and the Bear and "the people." Civilization, or rationality, marched into that natural world in the form of the Sartorises, and then the Sutpens and Compsons, who brought with them their laws and pianos and grandfather clocks and slaves—the original sin—and because they were dependent on their slaves, their dependence brought the war and the Reconstruction, and then the shadow of slavery called segregation, and no redemption for the sin. The fever had broken, but the virus remained in the blood.

When the shadow of the sash appeared on the curtains it was between seven and eight o'clock and then I was in time again, hearing the watch. It was Grandfather's and when Father gave it to me he said I give you the mausoleum of all hope and desire. . . . I give it to you not that you may remember time, but that you might forget it now and then for a moment and not spend all your breath trying to conquer it.

And so, even into the 1920s, the sins were repeated, a "redeemed" white man had yet to be born, and only the lower forms of Southern white humanity, the Jason Compsons and the proliferating Snopeses, could slip undetected under the moral screen, like mink or weasels. And those reappearing blooms of Southern manhood born with the grace, strength, or sensitivity to lead, like Sutpen's quarter-Haitian son, Charles Bon, or Joe Christmas, or Quentin Compson, were doomed.

In other words, human time, what we'd call history, is polluted. And natural time is not available to whites. From Faulkner's perspective, time has a moral dimension, and only those who had been spared the immorality of slave ownership escape the temporal judgment of cyclical return. Dilsey, the Negro servant, is the true mother of the Compson clan. Only she can settle the howling Benjy and master the moments of marriage and burial. Only she can turn aside Jason. The famous words Faulkner applied to the Negroes of Mississippi, praise for their "endurance," and that they "prevailed," are temporal judgments of the highest order. Their historic victimization rendered them in our terms timeless, but in the jazzman's language, "in time."

(Faulkner should not be held to task. Finally, we don't care if

our hired mothers are racially, culturally, or linguistically different from ourselves, so long as they are demonstrably separated
from us in time. We don't want our nannies and housekeepers to
show up wearing a Walkman and asking about cable programs;
we want them to have "endured" or "prevailed" from an earlier
era, where, presumably, maternal values were more central, less
compromised, than in our own.)

In the great modernist novels, *The Magic Mountain, To the
Lighthouse, Ulysses,* in most of Lawrence, most of Faulkner, in
Proust, in Gertrude Stein, time is manipulated in order to keep
the moral issues alive. In film, certainly the most extreme and
most successful example of temporal manipulation in all the arts,
the present moment is eternal. Cinema's narrative compression
suppresses time altogether. The filmgoer creates the temporal
dimension, just as the gallery visitor mixes the impressionist
palette in his head. All is surface, which is not to say they lack
depth. They get where they're going without time.

Is there a moral component to time? Faulkner certainly
thought so. Günter Grass *knows* there is. There is Southern time,
German time, African time, Irish time, Latin American time, all
of them set to a different moral, that is, aesthetic, ideal. Contemporary Latin American novelists like Juan Rulfo, Carlos Fuentes,
García Márquez, and Vargas Llosa have taken Faulknerian time,
Proustian time, the Catholic calendar, and even a bit of pre-
Columbian aboriginal time to keep history alive for the creation
of their own sophisticated myths of eternal return. For writers
of my generation, time as a subject, a dramatic event in itself
worthy of a new form of telling, is forever associated with reading Borges for the first time.

The world was made conscious of Protestant time and the so-
called Protestant work ethic back in the nineteenth century.
Protestant travelers and imperialists imposed their hyperconscious temporality wherever their influence spread. The Protes-

tant clock, not the Catholic cross, was their god. To be prompt and reliable was the surest outer sign of an orderly and responsible inner life. They also reported back on their unsuccessful temporal conversions, the slackness of Catholics and Muslims and Hindus, of "native" people just about anywhere. The *mañana* mentality was a childlike moral failing.

"Cultural time" obviously differs from the clock, the calendar, or the civil day. In the same way that many contemporary Americans and Europeans live speeded-up lives, dipping in and out of many time zones in a single day, living only marginally within their own local time, there are other cultures that achieve the same freedom from local time by discounting it altogether. Robert Levine's *The Geography of Time,* the same book that introduced the resonant phrase "temporal millionaire," measures time consciousness across many societies, taking into account comparative estimates of elapsed time, chronological projections of likely time investment, degrees of reliability and tardiness, time obsession and time laxity—even pedestrian foot speed. In Brazil, Levine discovered a modern country that manages to live in "natural" time while observing the rituals of standardization. Business and legal hours are posted, class times are published, but only on special occasions are they expected to be observed. Every citizen is his own prime meridian.

And as for German time, here is Günter Grass, speaking in 1982:

Everything that has thus far become a book for me has been subservient to time or has chafed under it. As a contemporary, I have written against the passage of time. The past made me throw it in the path of the present to make the present stumble. The future could only be understood on the basis of past made present. First and foremost, I found myself harnessed to German time, constrained to steer a course cutting obliquely across the epochs, disregarding the convenience of chronology. Epic moraines had to be cleared away, reality sloughed off again and again. There's no end to it. So many dead. And everywhere, even where life might release joy, and pleasure might take its fling, the great crime casts its shadow, which time cannot efface.

The effect of standard time, that is to say "reason," on a non-Western culture has been explicitly captured in one novel, Chinua Achebe's *Arrow of God*. Achebe's classic, in temporal terms, is a violent, deicidic clash between natural and rational gods, told in terms of colonialism and religion. Within the traditional Ibo religion, the priestly interpretation of the lunar calendar determines the proper moment for the planting of the village yams. Miss it, and the village loses a month; lose a month, and the crop is doomed. The starving village becomes ripe for plunder and takeover. Because the British colonial authority has imprisoned a village priest on a minor charge (largely to teach him respect for their own power), he is not available to signal the moment for planting the yams. Because the planting cycle is missed, the village starves, the old Ibo gods are discredited, and Christians gain a foothold.

It's about time. It's all about time. A change in the pace of change.

William Butler Yeats was another of the time-obsessed. His theory of the gyres, great, cyclical collapses and rebirths of civilization, are perhaps the broadest statement we're likely to get, outside of Hinduism or quantum physics, of time consciousness. "Sailing to Byzantium" (1927) was published just a year before *The Sound and the Fury*, five years after *Ulysses*. The famous and often-quoted last stanza could have been whispered in the ear of Quentin Compson, or for anyone seeking release from the ticking clock. While addressing the familiar issues of nature and reason, it also anticipates the next stage of the conflict. Once nature is replaced by reason, and reason fractured by subjectivity, how do we recompose reality, and a sense of self?

> *Once out of nature I shall never take*
> *My bodily form from any natural thing,*
> *But such a form as Grecian goldsmiths make*
> *Of hammered gold and gold enamelling*
> *To keep a drowsy Emperor awake;*
> *Or set upon a golden bough to sing*
> *To lords and ladies of Byzantium*
> *Of what is past, or passing, or to come.*

In Fleming's eighty-eight years, the world redefined itself as profoundly as during any interval in human history. In their various ways, every significant figure or invention in the nineteenth century has been hailed as revolutionary. They did not build on previous knowledge or practice but, in effect, wiped them out. How did we move, lift, bore, before steam? How did we survive childhood before vaccination? What were newspapers good for before cable transmissions? The world could never go back, after Darwin, Pasteur, Edison, Seurat, Marx, James, Monet, van Gogh; or after steam engines, electric motors, motion pictures, X-rays, telephones, railroads, internal-combustion engines, wireless transmissions, photography; or after the birth of the natural and social sciences.

It is the privilege of each human generation to feel it has been born into the time of greatest speed, greatest confusion, greatest advance, greatest peril in all human history, and surely that generation is right. Claims for our era, however, fall a little short of those superlatives. My father drove his car as fast and as far in a day as I can; we flew (not as fast or as far, but with a compensating sense of luxury and adventure), we listened to radio and could visualize the scenes and characters more vividly than on any television show. Only the computer and its applications have altered the velocities of our life, and given definition to the past twenty years. Victorians still win the competition for determined stability in a world of change.

TRACING THE origins of modernism, picking out our ancestors from brittle, old class pictures, has become an intellectual vanity of our age. Each new biography, rediscovered painter, each retelling of fading events, each reconstruction of the mid- to late-nineteenth century pushes the frontiers of modernism back another year, or decade. The shifting temporal markers are like new archaeological finds, or new dinosaur bones, each new discovery pushing back the frontiers. And standing silently in the wings, rarely called on and barely introduced, are Fleming and Abbe and Allen and Dowd, and the giant issue of standard time.

When I entered college, modernism was thought to have been invented by the generation born in the 1880s, who came to maturity early in the twentieth century—Einstein, Joyce, and Picasso, and the major innovators in Paris dance and music. Roger Shattuck's *The Banquet Years* served as the handiest guide. Now modernism is already alive in 1880, created by figures in science and the arts born at mid-century or even earlier. We didn't know much about Vienna and Berlin, and America didn't count at all. Modernism had something to do with the techniques of cinema, and with the use of the telephone, with the popularity of Freudian analysis, and with relativity and the demographic shifts from country to city and ghetto to mainstream. It had to do with forty years of peace on the European continent, from about 1875 to 1915, and with the rising expectations of newly urbanized minorities in the Austro-Hungarian Empire. It had to do with many things, all vaguely synergistic, all loosely interrelated. And, of course, with a generalized sense of breakdown—sexual, religious, social, and political. But standard time was never mentioned.

When looked at closely, modernism never really conformed to any single convenient description. Discontinuity is a major factor, a rebellion against the Victorian notions of rounded connectedness, the "well-stuffed" ideal. The "art for art's sake" credo, the notion of the artist standing apart from society and owing it nothing, clearly derives from a rebellion against the Victorian notion of serving society and of being, above all things, useful. But that attitude is as much Flaubertian as Joycean, despite the gap of fifty years. Modernism has been pushed back another generation and a half, to the 1870s, traced (by Kern and Everdell, notably) through physics and mathematics. Modernism, at least in spirit, was there all along, in medicine, in poetry, in a series of remarkable inventions, but mainly in the new perception of accelerated speed, the challenge to the time-space continuum.

As I've tried to suggest, modernism is also about the cutting and faceting of time, breaking up continuity, or flow, in favor of emphatic shards, away from nature and toward abstraction. Everdell has settled on a seemingly simple but endlessly complex definition. Modernism, he suggests, might only be "a change in the pace of change." That statement contains much of modern science, for it assumes a universe in constant motion, and no fixed point to measure it from. It looks not at content but at the framing of content. Modernism is about speed and the expectation of even greater speed, and about the attempt to hold on to a fleeting familiarity before it slips away. What modernism replaced was slowness, or the natural.

Sir Sandford Fleming had passed from the scene when novelists and poets, half a century after the Decade of Time, began agonizing over its dominance, and ways to transcend it. He could have helped Quentin Compson and Nick Adams, and might have understood Yeats and Woolf. His concerns, after all, became their obsessions—how to live in the moment while keeping an eye on eternity. How to live, in short, in a duality of time. The fussy Victorian contrivance that was the Fleming watch-face, taken to its most abstract level, placed its user both in time and in space, both locally and globally, both alone and in community.

He was, finally, just a Victorian. He wanted to tame and to humanize the great adventure of his era, the new reach of speed, power, and distance. He could not have foreseen its legacy, an inheritance of alienation and loneliness.

WHAT WAS A Victorian? A moralist, a reformer, a scientific progressive, probably a racist (although, like Cleveland Abbe, they can pleasantly surprise us), a colonialist, and often an imperialist (but with the best of intentions, as they often qualified it), a religious secularist, an intense (even sentimental) nationalist, probably an unthinking and rather narrow-minded Protestant. (Even Abbe's tolerance was strained when it came to Catholicism.) Because of their confidence and security, they were grand and glorious forecasters—the future would be like the present, only better. As the always enthusiastic Abbe had written from Russia in 1866: "Every year the world seems to me to be growing smaller and smaller: steamliners across the oceans, Atlantic and Pacific and Mediterranean; railroads across America and Europe; telegraph from San Francisco to San Francisco all the way around. If we do not hitch onto the moon and quarry our granite there it won't be the fault of the Yankees."

These Victorians make us feel like visitors to Donald Barthelme's "Tolstoy Museum," weeping paper streamers from our eyes, weeping for the sheer scale of their achievement, weeping for the world that abandoned them, weeping for the heroic burden of their understandable vanity. So much accomplishment, so much confidence, such touching ignorance.

Fleming's greatest single achievement in later life, after time had been settled, was his single-minded oversight of the worldwide sub-Pacific cable. In 1902, from his home in Ottawa, he sent two messages to Australia: one via London, the other via Vancouver. The responses were received back in Ottawa some eight hours later after twin circumnavigations of the world, through Fiji and Sydney, then back through India, Egypt, Malta, Gibraltar, and London, a proudly "all-red" route never straying from British soil. It was a miraculous moment—around the world in eight hours (and perfectible to even fewer). It must have reminded him of the moment five years earlier, now infinitely expanded, when he'd sent a message from Ireland to London and received a response an hour later, eight miles down the tracks.

When the cable was finally operating, in 1902, it seemed a miracle, but it would soon be trumped by Marconi's wireless, leaping the Channel, and then the Atlantic within the decade. His great dream for the cable was imperial unity, the "real time" transmission of political "intelligence" (meaning information). He had shrunk the world in time and in space; all that was now left in his great project of reforming the world was to use the wondrous technologies of the new century to reduce the real costs of distance, and the psychological damage of isolation. To that end, he undertook the destruction of the British cable monopoly, the Eastern Extension Company, in the name of a fair price for the cost of Australian and New Zealand cable transmissions. He could not have foreseen that standard time would help loosen community cohesion and encourage, in time, a kind of rootless anomie. And never could he have imagined the breakup of his beloved British Empire just thirty years after his death.

The Prime(s) of Mr. Sandford Fleming

Prime meridian, prime meridian, I'm sick of prime meridian.
—FLEMING (Venice, 1881)

The application of science to the means of locomotion and to the instantaneous transmission of thought and speech have gradually contracted space and annihilated distance. The whole world is drawn into immediate neighborhood and near relationship, and we have now become sensible to inconveniences and to many disturbing influences in our reckoning of time utterly unknown and even unthought of a few generations back.
—FLEMING (1884)

VENICE, 1881; ROME, 1883

IN THE FOUR years between his CPR dismissal and the Prime Meridian Conference, Fleming was not merely "riding the gondolas" of Venice, as Pierre Berton charged. After 1879, when he no longer took his identity from engineering, or the Canadian Pacific Railway, he began working full-time on the two major projects of his life, world standard time, and the worldwide undersea cable. They were, of course, related endeavors—shrinking the world to the speed of new technology, one might say—although they involved distinctly different strategies and different casts of characters.

For standard time, he was writing and delivering papers as a member of the Metrological Society and as chairman of the time convention of the American Society of Civil Engineers. As the Canadian Institute's delegate to the World Geodesic Conferences in Venice (1881) and Rome (1883), he gave two papers of note that led directly to the calling of the Prime Meridian Conference in 1884.

Standard time, at least for the first few years before he took over a leadership role, seemed blessedly free of politics. His opponents were normally scientists and academics, from whom he could expect a certain level of civility. There were a few testy encounters with the irascibly territorial William F. Allen, but disagreements for the most part were conducted on a suitably elevated plane. And he was fortunate to count Cleveland Abbe as a friend and advisor, a man much like himself who preferred to manage from the wings rather than to take center stage. The exceptions, like the astronomer-royal George Airy, or the eccentric Piazzi Smyth, and even Fleming's mysteriously dyspeptic archenemy, the Canadian-born head of the U.S. Naval Observatory, Simon Newcomb, offered containable opposition.

At the World Geodesic Congress in Venice in 1881, fellow metrologists, particularly the president of the Metrological Society, Cleveland Abbe, and the president of Columbia University, Frederick Barnard, urged him to propose a conference of diplomats and astronomers to set a prime meridian for the world. President Barnard, a native New Englander then in his late seventies (and, not coincidentally, the figure for whom Barnard College was later named), is the man most responsible for women being admitted to a prestigious university as early as 1890 (the year after his death). His stature is a fair indication of the standard of rationality and passion for reform that the standard time movement attracted. Barnard had been president of the University of Mississippi until 1861, when, being a Union sympathizer, he'd left the South and taken on the presidency of an obscure classical college in New York City and transformed it, Prince Albert–fashion, into a modern university with associated technical colleges, particularly in his specialty of mining engineering. He was a vigorous and effective proponent of time reform, known throughout the world, but a generation older than Fleming, and limited to the role of advocate and theorist.

Cleveland Abbe, a Washington insider, knew that the United States would never initiate a proposal for such a conference. Fleming, for his part, realized that France would never attend a meeting in London. Thus, an intermediary's role opened for Canada (perhaps for the first, but certainly not the last, time). Fleming's personal prestige and the wide recognition he enjoyed, as well as his very active chairmanship of the Standard Time Convention of the American Society of Civil Engineers, permitted him, Canadian fashion, a kind of high-level entry to both the United States and Britain. At the 1883 Rome meeting, two years later, the general assembly adopted his motion left over from the Venice congress of 1881. Britain sent notices. President Arthur issued his invitations, and the meeting was formally set for Washington, D.C., to begin on October 1, 1884.

THE POLITICAL opposition to the worldwide cable, Fleming's final undertaking, was a far different matter from polite academic disagreement, or even political competition. The cable battles would insert Fleming into the front line of international power politics between Britain and the United States over Hawaii (he felt that the proper diplomacy might have snared Hawaii, or at least one of the chain, for Britain), with the Dutch in Indonesia, the Japanese, and the French in East and North Africa—wherever an undersea cable was forced to surface on foreign-held land. He entered cabinet-level conflicts in every

"white" British colony on every continent, supporting autonomous forces against the British-owned communications monopoly, the Eastern Extension Company, and its head, the media
baron Sir John Pender. Pender's monopoly controlled all cable
traffic between Europe and the South Pacific, through a network
of uncertain concessions won from a host of foreign governments, and set a crippling cost-per-word tariff to recoup a guaranteed 6-percent profit for its inefficient operations. As a result,
Australia and New Zealand suffered from a sense of isolation far
greater than their distance alone should have imposed.

The challenge to the power of Pender and that of his allies in
high places, many of whom were on his payroll, is quite extraordinary, particularly considering Fleming's preference to act as an
unassertive, behind-the-scenes facilitator. Something in Pender
brought out Fleming's confrontational streak, reawakening his
dislike for the English power structure, the corruption of monopoly capital, and the abuse of its colonial mandate. Simply to
get his proposals considered, Fleming was obliged to work with
agents and lobbyists in Washington, paying their fees from his
own pocket. It also brought out in Fleming a degree of political
commitment that had been missing in his life to date.

His two projects of world time and the world-circling cable,
taken together—considering their reach and implication, and the
sophisticated and varied approaches they required, the tenacity,
commitment, and costs, as well as the organized opposition he
encountered—are an extraordinary individual accomplishment.
But their undertaking is in keeping with his tragic vision of engineering: when they are completed, the world benefits, and their
authorship disappears.

OVER THE YEARS, from his first paper in 1876, he had simplified his time proposals to a core set of twenty, which he summarized at the time of the Venice meeting in 1881 (where, indeed,
he and a daughter rode the gondolas). The shadow of his first

Toronto paper still hung over the various revisions and improvements—the "cosmic" and "local" day, the system of lettering and numbering of the meridians. But there was no longer a mythical timekeeper in the center of the earth. Instead, there was reference to Z, the as-yet undetermined zero longitude, or the prime meridian. Although he was careful not to hint at his solution, he pointed out how embarrassing and unnecessary the multiplication of national prime meridians was:

> For a number of years the question of reducing this number has been under consideration. . . . The question has been examined by different societies, and various proposals have been submitted, but unanimity with respect to the selection of a prime meridian, to be common to all nations, has in no way been attained. Repeated efforts have been made to gain general concurrence to the adoption of one of the existing national meridians, but these proposals have tended to retard a settlement of the question by awakening national susceptibilities and thus creating a barrier difficult to remove.

Again, those delicate national susceptibilities! Too many primes, but no existing prime—including, presumably, Greenwich or Paris—would do.

The twenty proposals of the Venice paper are worth close attention. The first four state the obvious: there should be a world day. There should be a single meridian with the day to begin on that meridian, and the meridian should be established by the concurrence of all the world's twenty-six "civilized" (in this case, independent) nations. Proposal number five, however, states: "For reasons elsewhere given it is suggested that the prime meridian and time zero shall be established through the Pacific Ocean, entirely avoiding the land of any nationality." By the eighth proposal, delegates to the Venice Geographic Conference were asked to approve another item on the Fleming agenda. The twenty-four time zones ("fifteen degrees, or one hour distant from each other") were to be named according to the English alphabet, omitting the letters I and U. (He had not got around to eliminating U altogether himself, since his sample charts, at least those used in Venice, still employed it.) There were to be two kinds of standard time for the world, "cosmic" and "local."

Proposal number nine reads:

> The unit measure of time, determined as above, shall be held to be a *day absolute,* and irrespective of the periods of light and darkness which vary with the longitude, to be common to the whole world for all non-local purposes. To distinguish it from ordinary local days, this space of time may be known as the "Cosmopolitan" or "Cosmic Day." The hours, minutes and seconds of the cosmic day, and the days themselves, may be distinguished by the general term, *cosmic time.*

Cosmic time, in Fleming's scheme, was to become the time of science and global communications. The absolute, or cosmic day, the twenty-four hours beginning and ending on the yet-to-be designated prime, would necessarily contradict local times. The cosmic noon, for example, would be any other hour of the day or night at any given location on the globe. "To promote exactness," Fleming wrote, "it may be employed in astronomy, navigation, meteorology, and in connection with synchronous observations in all parts of the world. It may be regarded as the time which would be used in ocean telegraphy and in all operations of a general or non-local character." The general public would be spared much contact with cosmic time, except as it touched on cable communications or train travel. The cosmic clock was for professional use, meant to unify the naval and astronomical days (more on that later), and to accommodate meteorologists, like Cleveland Abbe, toiling over the hourly updates of their isothermic and isobaric charts.

Fleming's apparently casual mention of "ocean telegraphy" is a hidden reference to the proposed Pacific Ocean prime meridian. Without the technology and authority of a modern observatory anywhere near his proposed watery prime, Fleming would be forced to rely on telegraphy from the nearest outpost of civilization (Auckland? Honolulu?) to provide precise time for the entire world. Canadians were well advanced in the use of telegraphy, as were the French and many other European countries, most of whom, by 1881, had standardized their national times to their main national observatories. In the United States, accurate time was sent by signal from observatories to the Western Union Company and the railroads, which in turn sold time signals, usually in the form of time-balls, to cities, industry, and individuals. Time had become a profitable business for otherwise cash-starved observatories, and, of course, for Western Union. (A year's subscription to the Washington time signal could cost, depending upon the distance from the signal's source, upwards of $500.)

Time as a commodity was taken very seriously indeed. The Harvard College Observatory, which sent its time signal throughout New England and to the Western Union Company, built its telegraphy room with all the sanitary and climatic controls of a modern silicon-chip room. Leonard Waldo, director of time services for the observatory, and yet another member of the Metrological Society, reported that the instrumentation and telegraphy room was "nearly as practicable free from changes of temperature, with a floor of sheet lead, and walls filled with dry sand, the doors having felt linings." With such attention to precision having to be taken in the relatively benign precincts of Cambridge (those telegraph lines extended only to Boston, across the Charles River), it does raise serious doubts that any kind of reliable signal could be sent to, or received from, an exposed device in the South Pacific. Thus Fleming's Pacific prime proposal glossed over an important objection. The only way of placating

national jealousies was to deny the Z meridian local access to sophisticated astronomical equipment. It sharpened the delegates'
calculation between the easy choice of Greenwich and the riskier
adventure of its nether-arc.

There is, of course, a yet larger calculation that Fleming had
been hinting at all along. In his earliest Toronto paper, with its
mythical clock in the center of the earth, or (in theory) hovering
in the clouds over it, there was always the suggestion that *any*
prime will do, so long as we free ourselves from a social and historical dependence on local time. The Victorian utopian streak
runs throughout Fleming's proposals, along with the sense that a
single universal day is eventually all that is required by an enlightened world citizenry. Local time, like Marx's state, would
eventually fade away.

Fleming's friends, the academicians and delegates to the 1881
World Geographic Conference in Venice, were well-acquainted
with the accuracy and flexibility of telegraphy. For the French,
the perfect deployment of technology could serve as a potential
anti-Greenwich argument. For scientists in general, electric signaling was the future. It was progressive, it transcended borders,
and it was instantaneous. But the conference delegates were political innocents. No one properly foresaw the ingenious objections to remote signaling that would be raised, three years later,
by well-trained diplomats representing narrow national interests.

Proposal eleven of Fleming's twenty returned to dual-track
time: "cosmic time" denoted by letters, "local time" by numbers.
Proposal twelve spelled it out: "Cosmic time shall be so lettered
that the hours will correspond with the twenty-four standard
time meridians. When the sun passes meridians G or N it will be
G or N time of the cosmic day. When it becomes Z time, that is to
say, when the (mean) sun passes the zero meridian, at that moment, one cosmic day will end and another begin."

Proposals thirteen to sixteen concerned the definitions of
local time, and extended assurances that for the vast majority of

world citizens going about their daily business, there would be no disruption of accustomed time standards. The seventy-five time standards of North America would shrink to four, but that process was well advanced, and generally supported anyway. There would be twenty-four standards of time for the world, lettered appropriately. "It is intended that local time at any place on the surface of the globe shall generally be regulated by the standard meridian nearest or most convenient to such place in longitude," read proposal number fourteen.

Fifteen and sixteen went on to designate the naming of the zones by letter to correspond with the letter of the nearest meridian. (By "standard meridian" was meant the fifteen-degree meridians that signaled an hour's time change.) Rather than living in Pacific or Mountain time, one would be obliged to adjust one's horizons to the infinitely less pictorial U or T time. Fleming never surrendered his distaste for "local times," and the chauvinism they engendered. If time for the whole world was regulated by letters instead of descriptive designations, local times (assuming we had an agreed-upon Z) would be drained of their local associations.

Proposal seventeen has Canada written all over it: "It is proposed that standard time shall be determined and disseminated under Government authority; that time signal stations be established at important centres for the purpose of disseminating correct time with precision, and that all the railway and local public clocks be controlled electrically from the public time stations, or otherwise kept in perfect agreement." He saw time as a free, common resource, not as a privately held property on the order of the American railroads or the sold time signals from the Western Union Company. Fleming was always a government man, deeply suspicious of the motives of capital and profit, a conviction that only grew in coming years. His last great battle with capital, begun in his mid-eighties, saw him launch an inquiry against Canada Cement, a conglomerate that had named him honorary director but whose president and inner circle had enriched themselves, he charged, by watering the stock at the first public offering.

With Fleming's last three proposals, we enter a mirror-world to every earlier, Greenwich-prime assumption. Since he was calculating from a Pacific (not a Greenwich) Z, moving eastward, not westward, the Y, X, W, and V meridians (that is, the 165th, 150th, 135th, and 120th longitudes) all fell over the open Pacific. North America began at the 105th degree of west longitude of California and British Columbia, not with the more customary 60th degree of Newfoundland. California and the Pacific-rim states and provinces thus made their debut as the landfall—the first, not the forgotten, entities.

Proposals eighteen, nineteen, and twenty concerned the application of the system to North America. There would be four standards, U, T, S, and R, running from west to east. Standard U would extend eastward to Idaho, Utah, Arizona, and Nevada. The next standard, T, would include all of Mexico, and the Plains states and provinces as far east as Kansas, the Dakotas, Texas, and Manitoba. Standard S covered all states on both sides of the Mississippi River, plus Michigan, and R kept time for everyone else, all the way to Nova Scotia.

Fleming summarized his scheme, however turned-around it might appear to us (and doubtless, to some of the East Coast and European delegates in Venice), with his usual insouciance:

> The foregoing is a general outline of the proposition. It must be evident that the system of cosmopolitan time would be a ready means of meeting the difficulties to which I have referred. It would render it predictable to secure uniformity, great simplicity, perfect accuracy, and complete harmony. The times of places widely differing in longitude would differ only by entire hours. In all other respects standard time in every longitude and latitude would be in perfect agreement. In the-

ory every clock in the world would indicate some one of the twenty-four hours at the same instant, and there would be perfect synchronism with the minutes and seconds everywhere around the globe.

By the system proposed, instead of an infinite and confusing number of local days, following the sun during each diurnal revolution of the earth, we should have twenty-four well-defined local days only; each local day would have a fixed relation to the others, and all would be governed by the position of the sun in respect to the prime meridian. Those twenty-four local days would succeed each other at intervals of one hour during each successive diurnal revolution of the globe. The day of each locality would be known by the letter or other designation of its standard meridian, and the general confusion and ambiguity which I have set forth as the consequence of the present system would cease to exist.

His proposals were approved, and became the basis of the World Geodesic Congress's recommendation for action. Cleveland Abbe, Frederick Barnard, and Fleming raised the issue of a Prime Meridian Conference to settle the issue of a prime once and for all. The delegates assented but delayed until their next meeting, two years hence in Rome, thus allowing their well-placed officers to make approaches to the American government. Fleming, following Abbe's advice, immediately set to work with personal memos and speeches to American chambers of commerce, railroad conventions, and shipping and insurance companies, as well as with more formal approaches through the governor-general and the British Colonial Office. His proposals closed with a signature flourish: the demand for the twenty-four-hour clock, the eradication of that temporal anomaly that had got him started five years earlier.

By 1883 North American railroads had standardized their times according to Mr. Allen's designs, and on the Continent the first service between Paris and Constantinople had been launched. Time was in the air, and on the ground, and for the first time in history a coherent system for regulating it had been proposed. Only a final meeting stood between a Fleming-inspired standard-time system for the world and what appeared to be its foregone diplomatic ratification in Washington, and that was the 1883 Rome meeting of the Geodesic Congress.

FLEMING'S EARLIER papers, particularly his presentations at the Canadian Institute in January and February 1879, "Time-reckoning and the Establishment of a Prime Meridian," had been forwarded to eighteen countries by the Marquess of Lorne (the governor-general), and to London for review and general circulation. There, they had met with the astronomer-royal's disapproval. In 1881, Airy would not be alive for the next stage of the battle, and his successor as astronomer-royal fully endorsed the standard-time reform.

Defections from prominent astronomers must have been painful to Fleming, especially that of Simon Newcomb, chief of the U.S. Naval Observatory, and in later years, author of a still-useful astronomy text. Newcomb was something of a self-taught mathematical genius. He was also a notoriously difficult and rather eccentric individual, but a close friend to his fellow Washington astronomer, that genuinely pleasant soul, Cleveland Abbe. When Fleming had sent out a questionnaire on the utility of standard time to the membership of the American Society of Civil Engineers, he kept the dozens of responses neatly bundled in their appropriate envelopes—all except those of the astronomer Simon Newcomb. To the third question, "Do you consider it advisable to secure a time system for this country which would commend itself to other nations and be adopted by them ultimately?" Newcomb had responded: "No! We don't care for other nations; we can't help them, and they can't help us."

Fleming's second question which read, "Do you favour the

idea . . . of bringing the Standards of Time of all countries into agreement?" got an even more unwelcome response: "See no more reason for considering Europe in the matter than for considering the inhabitants of the planet Mars." The opposition of Newcomb worried Fleming sufficiently that he felt obliged to warn others, with uncharacteristic violence (but always inside quotation marks), of his unnamed presence as "a niggar in the fence" (sic), a hidden, treacherous obstruction to any scheme for world standardization. Other dissenting astronomers saw the drawing of a prime meridian, if it had any virtue at all, as an opportunity to right historical and religious wrongs: Piazzi Smyth, of course, plumped again for the Great Pyramid of Giza. Others spoke up for Pisa, to honor Galileo, Jerusalem, or the Azore Islands, the original European prime. Everyone realized, however, that there were only three true contenders: Greenwich, Paris, and Fleming's "nether arc."

Fleming had his reliable supporters from the astronomy fraternity, like Otto Struve of Russia, and the Spanish, Italian, and Mexican astronomers who were delegates in Rome and would be in Washington as well. At the Rome conference of 1883, Fleming reiterated many of his older ideas, but this time pushed harder for an anti-Greenwich prime, that is, the Pacific Z meridian.

Resolutions three, four, five, and six at the Rome meeting concerned the prime, and the vote could not have been reassuring for Fleming. Greenwich, on the basis of the popularity of its charts and the number of countries already employing it, won the vote simply on grounds of sheer convenience. Fleming's anti-prime was the choice of many, like Struve and Juan Pastorín of Spain, and was judged an acceptable second choice should Greenwich, for some reason—"national susceptibilities" being the obvious—not be adopted at the upcoming Washington conference. In the fourth resolution, delegates voted to count the meridians with consecutive numbers in a single direction, zero to 360, not splitting the earth into 180 degrees of east and west longitudes. This, too, was Fleming-inspired. Double-counting the longitudes was as distasteful to his sense of time's "flow" as double-counting the hours of the day.

Resolutions five and six concerned the universal day, and handed Fleming a curious victory. The universal day would start at midnight on the Pacific anti-prime, that is, when it was noon (of the previous day) at Greenwich. This would permit the unification of the astronomer's professional day with that of the universal day (astronomers traditionally counted their day from noon till the following noon in order to preserve a single calendar date for their nighttime observations.)

All exceptions to standardization, the various "professional days," would eventually have to fall. Astronomers could be counted on to resist, just as admirals had when the nautical day had been eliminated in Nelson's time. Both professions held themselves apart from the midnight start of the civil day. (The traditional nautical day had also run from noon to noon, which retarded many naval communications by a whole calendar day, with catastrophic results in history.) The nautical day had already been regularized in most countries, but without an internationally binding protocol, hot spots of resistance might flare. Fleming's dual-track time, the elaborate system of letters and numbers, and his imaginative responses to anticipated French objections to Greenwich, were under serious threat.

OCTOBER 1, 1884; WASHINGTON, D.C.

VIEWED FROM A distance of a century and a quarter, the Prime Meridian Conference seems thoroughly contemporary in its mixture of good science and bad politics, and doomed to failure by the inevitable friction between their fundamental cultures. The subject to be decided might seem more suited to a philosophy seminar than a crowded hall hazy with cigar smoke. Put simply, it was to decide where time begins, and the proper way to measure it.

Once the conference was under way, however, politics—in its worst sense—took a leading role. Science establishes bedrock methods and principles that cannot be compromised, knowing that the fudging of science is the destruction of reason. Diplomacy fashions a world of elaborated compromise, transforming differences of language, culture, race, and religion into the formulas of treaty and protocol, knowing that the alternative to negotiation is the breakdown of civilization. Only a subject as broad as time could even begin to bring such diverse positions together.

There were only twenty-five "civilized" (independent) countries recognized by the United States in 1884, one in Asia (Japan) and one in Africa (Liberia), the rest in South America, Western Europe, and the Caribbean, plus Hawaii and Russia. When President Arthur sent out invitations in December 1883, all twenty-five answered the call. Ten months later, however, due to cholera quarantines affecting the Mediterranean countries, only nineteen nations were able to make the opening. Five of the thirty-five delegates present represented the host country, four were from Britain (including Mr. Fleming, an "honorary" member from the unrecognized Dominion of Canada), two from France, three from Russia. The representatives of Turkey and Japan, an ambassador and an astronomer respectively, were the only non-Christians. Dual or multiple representation usually indicated the presence of an astronomer, along with the sitting Washington ambassador. When, by the third session, all late-comers were seated, it was apparent that only Denmark did not bother to attend. Fleming, always the multi-tasker, left Ottawa ten days before the opening in order to attend board meetings in Montreal. Ever meticulous, he made note of his finances ($212 on hand), of his accommodations (the Biggs Hotel), and of his attending Negro church services with Professor Abbe. He noted the first day's temperature as well: ninety degrees, following the hottest and driest summer on record.

President Arthur's letter, though formal and formulaic, was to serve as a general guide, in good diplomatic fashion, through coming difficulties:

> In the absence of a common and accepted standard for the computation of time for other than astronomical purposes, embarrassments are experienced in the ordinary affairs of modern commerce; that this embarrassment is especially felt since the extension of telegraphic and railway communications has joined states and continents possessing independent and widely separated meridional standards of time; that the subject of a common meridian has been for several years past discussed in this country and in Europe by commercial and scientific bodies, and the need of a general agreement upon a single standard recognized; and that in recent European conferences especially, favor was shown in the suggestion that, as the United States possesses the greatest longitudinal extension of any country traversed by railway and telegraphy lines, the initiatory measures for holding an international convention to consider so important a subject should be taken by the United States Government.

President Arthur's letter manages to indicate that the conclusions of preparatory conferences, such as Rome and Venice, were to be considered. The link (the "embarrassment") between the speed of railway and telegraph communications and the confusions of national primes is noted. And if the president's letter was not specific enough, the group was soon to be addressed by the secretary of state, Frederick Frelinghuysen, who set the agenda succinctly:

It gives me pleasure, in the name of the President of the United States, to welcome you to this Congress, where most of the nations of the earth are represented. You have met to discuss and consider the important question of a prime meridian for all nations. *It will rest with you to give a definite result to the preparatory labors of other scientific associations and special congresses, and thus make those labors available.* [Italics added.]

The phrases "definite result" and "make those labors available" are loaded indeed, and would come to play a definitive role in the next three weeks. Results, not further deliberations, were expected. The obscurities of science were to be made "available"— that is, legally binding. The Washington conference was to be a diplomatic, not a scientific, show.

Secretary Frelinghuysen's opening remarks commending the findings of "other scientific associations and special congresses" practically enthroned Greenwich before a vote could even be taken. Those congresses, especially the 1883 meeting in Rome, had already settled on Greenwich as the most logical and least disruptive choice, while regarding Fleming's anti-prime as a not unreasonable alternative. American railroads, whose new standardization ran on Greenwich-based time zones, had already threatened a strike if anything but Greenwich were chosen. But officially—that is, diplomatically—Greenwich was just another national prime, equal but not superior to ten others.

If Greenwich prevailed, ten proud astronomical traditions, along with their charts and maps, would have to be scrapped. Nine of the countries might go gently, but there was one that could be counted on to resist any assault on the dignity of its *ligne sacrée,* the Paris meridian.

LACKING AFFILIATION with a member country, Fleming had been accredited to the British delegation. His fellow British delegates included Professor Adams of Cambridge, the 1845 "plotter-by-planetary-perturbation" of Neptune; General Strachey of the Indian Army and the Council of India, who had hosted the 1869 eclipse party to South India; and Captain Sir Frederick Evans, the head of Britain's Naval Observatory. Cleveland Abbe and the secretary of the American Railroad Association, William F. Allen, were among the five American delegates. The others were Admiral Christopher Rodgers from the Naval Observatory, Commander Sampson of the navy, and the astrophysicist and spectroscopist Lewis Rutherfurd, another veteran of the Indian eclipse. Rear Admiral William T. Sampson and Rutherfurd, in particular, were formidably articulate, aggressive, and well-prepared.

The impressively white-bearded Admiral Rodgers was nearing retirement age, but Commander Sampson's glory days were still before him. He would enter American military history as the victor of the Battle of San Juan Harbor during the Spanish-American War, the man most responsible for the acquisition of Puerto Rico for the United States. He would even model men's suits in mail-order catalogues, an early instance of military-industrial cooperation. The two French delegates were the ambassador, Monsieur Lefaivre, and Jules-César Janssen, the world's leading spectroscopist and the founder-director of the Meudon Observatory in Paris. Many of the scientists were old friends, well known to each other from eclipse parties and various professional congresses. Never, however, had they been asked to represent the interests of their countries.

Janssen is an especially attractive figure, although the role he was asked to play in Washington would cast him, in the popular press, as a pig-headed obstructionist. Originally a musician from a struggling family, and with no more formal education than Fleming, he had trained himself in ophthalmology, writing a thesis on the effect of the sun's rays upon the cornea. Rather than set himself up as an eye doctor, however, he had turned his attention to the rays themselves, devising the most sophisticated ways of capturing and analyzing light from the sun and other stars. His 1869 photos were classics in their day and are cited and studied even to this day.

Based on votes at the earlier geodesic gatherings in Rome and Venice, and at the Montreal convention of the American Civil Engineers, Fleming had reason to believe that his efforts of the past seven years in research, writing, and lobbying might possibly result in the acceptance of his modified time proposals. Alone among the "Anglo-Saxon" delegates, he was sensitive to French objections. He understood that an overtly British-American solution to the prime-meridian dilemma would merely antagonize the French and result in the loss of the desired consensus. French intellectual prestige among other "Latin" countries might well drag many South American and some European countries with her. With such complications in mind, Fleming had nurtured Italian, Spanish, Russian, and Belgian support. The delegate from Spain, Juan Pastorín, the head of his country's observatory, was also the Spanish translator of Fleming's papers.

But at his most realistic, Fleming must have recognized that the tides of history were shifting against him. He had been stronger in Montreal, at the American Society of Civil Engineers' first foreign convention in 1881, than with the geographers in Venice a year later, and more in command at Venice than in Rome, two years after that. His standing was higher among engineers than among astronomers, and higher with astronomers than with the unknown world of naval officers and foreign diplomats. And now that "convenience" was being cited as an important, perhaps even a decisive, consideration over, say, fairness, or international harmony, the appeal of Greenwich might well be insurmountable.

By the end of 1883, North American railroads had standardized their time to Greenwich, but not to the convenient standards of what we'd recognize today as time-zones. Europe was linked by rail from Iberia to Istanbul, yet it still had no coherent time standard across its expanse. Britain, France, Sweden, and Switzerland all recognized single national times, determined by their national observatories, yet their times were not coordinated with one another. Germany, which observed five official times, had already indicated, through the last public statement of Field Marshal Helmuth von Moltke (head of the chiefs of staff of the Prussian army), its endorsement for a single Berlin time, if for no other reason than military preparedness. The governments of Italy and Spain—countries whose maritime fleets adhered to Rome and Cádiz prime meridians—were in significant agreement with Fleming's proposals for world standard time and a Pacific prime. Otto Struve already plotted some Russian survey maps from Greenwich charts, and was an enthusiastic supporter of Fleming's modified Greenwich scheme. In other words, parts of the world had drawn closer, but the remaining differences in some ways had hardened.

If Greenwich, or any variation of Greenwich, were adopted, Paris, Berlin, Bern, Uppsala, St. Petersburg, Rome, and Cádiz, among others, would lose their prime meridians, their astronomical charts, and their proud histories. The national astronomers from "lesser" traditions were, in effect, being invited to prepare their professional suicide.

THE CONFERENCE opened at noon with Frelinghuysen's welcoming speech and a brief formal gathering with President Arthur. An hour later the thirty-five delegates settled around the large oval table in the Diplomatic Hall and unanimously elected Admiral Rodgers president of the congress. Rodgers opened by assuring the gathering that the United States, despite occupying one hundred degrees of longitude, not counting the Aleutian Islands (which spanned the European equivalent of London to mid-Siberia), and twelve thousand miles of coastline, had no in-tention of urging an American meridian upon the convention. The United States had experimented with a Washington meridian in its first half-century, but had voluntarily surrendered it in favor of Greenwich in 1849. He alluded only indirectly to the contentious issue of choosing a prime meridian, focusing instead on the problem of prime proliferation:

> In my own profession, that of a seaman, the embarrassment arising from the many prime meridians now in use is very conspicuous, and in the valuable interchange of longitudes by passing ships at sea, often difficult and hurried, sometimes only possible by figures written on a blackboard, much confusion arises, and at times grave danger. In the use of charts, too, this trouble is also annoying, and to us who live upon the sea a common prime meridian will be a great advantage.

The French diplomat Monsieur Lefaivre responded immediately, if somewhat obliquely. He suggested that all motions and addresses be translated into French. The French demand was predictable, but it, too, was a kind of code. The admiral had strayed just a little too close to the third rail of nineteenth-century international politics, the French-English rivalry, otherwise known as "the susceptibility." The dread words "common prime," coming in the English language from a naval officer in an English-speaking country were not, to the French sensibility, scientifically "neutral." They were code words for Greenwich, a dagger in the heart of Paris.

The first session ended with notification by the other American naval officer, Commander Sampson, that the next day's session would raise two peripheral matters: first, the desirability of opening all meetings to the public; and, second, that of inviting commentary from eminent specialists who happened to be passing through Washington, or in residence. Monsieur Lefaivre served notice that he would oppose both proposals. Admiral

Rodgers pledged State Department support in finding bilingual secretaries to prepare French and English transcripts of each day's proceedings. In this, the State Department's efforts were to fail. It was the British delegate representing the Dominion of Canada, Mr. Fleming, who would discover bilingual secretaries in Victorian Washington. The delegates broke for tea, cigars, dinner, and drinks, and a visit to the telegraph office to gather instructions for the next day's session. Mr. Fleming repaired to his rooms at the Biggs Hotel and on Sunday joined his friend Cleveland Abbe for services at a Negro church.

True to Commander Sampson's promise, the second session opened on the two motions for open seating and distinguished visitors, and both were defeated. Monsieur Lefaivre pointed out that although the conference was partially scientific, for which visiting experts and public attendance were entirely appropriate, the conference was also diplomatic. To admit the public to deliberations that were by their very nature privileged would expose the conference to popular pressure by uninformed partisans. The same could be said of the participation of unaccredited experts—all of whom just happened to be major American and British scientists—whatever their eminence. French logic carried the vote. Even Great Britain and the United States voted with the French in rejecting it.

All was going well indeed for France until Professor Rutherfurd moved: "*Resolved:* That the Conference proposes to the Governments represented the adoption as a standard meridian that of Greenwich, passing through the center of the transit instrument at the Observatory of Greenwich." Here it was, the perfidious resolution that France had hoped to delay, perhaps even bury, brought up as the first serious resolution of the conference.

The sweet logic of diplomacy, as practiced by Monsieur Lefaivre, was about to enter a virtuoso phase. First, he stated, the motion was out of order. Since the conference did not constitute a deliberative congress, they were merely gathering information for some future congress with executive powers. He denied that the scientific findings of Rome had any application to the diplomacy of Washington. To this, Admiral Rodgers ruled that the motion was indeed in order, and was in fact the best way to stimulate discussion on the very issue they'd been brought to Washington to resolve. Lefaivre then appealed to the self-regard of the delegates: "This Conference is composed of various elements, among which are scientists of the highest standing, but also functionaries of high rank, who are not familiar with scientific subjects, and who are charged with an examination of this question from a political stand-point. It is, moreover, our privilege to be philosophers and cosmopolitans, and to contemplate the interests of mankind not only for the present, but for the most distant future."

At that point, Lefaivre turned to his scientific colleague, Jules-César Janssen, director of the National Observatory, to pursue the objection. And so, a strange spectacle unfolded. Two of the leading scientists from the same field, spectroscopy, friends and colleagues who'd been at the 1869 South India eclipse together, found themselves cast as opponents, not collaborators, representing their countries, not a common cause. Janssen steered the discussion back to first principles. "We have inverted the process," he began, "nominating a meridian before discussing the nature of a prime, or indeed, if a prime for all nations is necessary. Since we are not empowered to select a prime, but only to report on the deliberations leading to recommendations, the Rutherfurd motion is out of order."

Commander Sampson rose to the challenge and amended the resolution: "*Resolved:* it is desirable to adopt a single prime for all nations." And so, the old issue of a prime was resurrected. A single prime for all nations was indeed desirable. General Strachey quoted Frelinghuysen's welcoming speech. Rutherfurd quoted

President Arthur; it was foregone for both that a single prime was desirable, and that it was a waste of time to discuss it. Janssen got his answer in the form of a unanimous support for a prime.

Janssen rose to the occasion, in his country's finest tradition of imperturbable logic. With the Greenwich motion now on the table, he sought to end the conference with a preemptive strike. We have done our job, the principle of a prime has been upheld. We are not empowered by our governments to execute any other decision. There would have to be a second (and a third, and a fourth) conference to debate the actual meridian.

And this is where Rutherfurd's preparation saved the conference (and the world) vast time and annoyance. He quoted from President Arthur's follow-up letter to each foreign minister, including the French, which clearly stated the objectives of the conference. Acceptance of the invitation implied acceptance of the objectives of the conference.

Diplomatic meetings are a delicate interplay of wit and logic, bluster and modesty. France had just been unanimously defeated, the United States had mounted a brilliant defense, but Rutherfurd pushed his advantage just a bit too hard. The whole object of the conference was to fix a prime meridian; there must be some misapprehension on the part of the learned gentleman from France in thinking that this conference did not have the power to fix a prime meridian. It seemed to him that the delegates undoubtedly were ready to hear and express arguments pro and con in regard to that very question, and he supposed that every delegate had studied the matter before coming. He did not think that any delegate would be likely to have come unless he knew, or thought he knew, something about the matter.

All of which unleashed a flurry of denials from Spain, Sweden, Russia, Germany, Mexico, Brazil, and even Britain. None of them, they averred, was empowered to fix a prime—merely to recommend.

The final speaker of the second session was Sandford Fleming. He called attention to the language of the act of the U.S. Congress that had authorized the conference:

> ... [that] The President of the United States be authorized and requested to extend to the Governments of all nations in diplomatic relations with our own an invitation to appoint delegates to meet delegates from the United States in the city of Washington, at such time as he may see fit to designate, for the purpose of fixing upon a meridian proper to be employed as a common zero of longitude and standard of time-reckoning throughout the globe.

The second session ended, the Greenwich motion still in play.

In the corridor outside the doors of the closed session, Monsieur Lefaivre swore to Commander Sampson, "France will never agree to emblazon on her charts 'degrees west or east of Greenwich'!" It was a promise, against all apparent odds, that could have been taken to the bank. France had dodged the first bullet, but the Greenwich guns would not go away. There were five days before the next session. During the break, lacking a country, and diplomatic instructions, Fleming returned to Montreal for a board meeting of the Hudson's Bay Company.

THE THIRD session witnessed a new French strategy, that of demanding a "neutral" prime meridian. France denied favoring a Paris prime; but she was equally opposed to any prime that carried historic importance, or conferred national advantage. If Greenwich carried the vote, British astronomers would not have to change their charts; their traditions would continue into the next century, while everyone else's would die. France would be stripped of everything, without compensation. As Janssen put it, "Whatever we may do, the common prime meridian will always be a crown to which there will be a hundred pretenders. Let us place the crown on the brow of science, and all will bow before it." By "crowning science" he meant there should be no local commercial benefit attached to the designation. By "neutral" France meant that any meridian selected had to be culturally naked and not fall on the continents of Europe or America. Only an ocean-based prime, like Fleming's anti-prime, met the standards. Fleming and France became unstable allies, one from a desire to implement standard time without dissension; the other to sidetrack the debate for as long as possible.

Arguing against neutrality, or even attempting to define it, is philosophically treacherous. (It is not the sort of argument that British and Americans make easily anyway, and it led, over the next two sessions, to flashes of anger on both sides. "Gentlemen, I remind you, we are not all belligerents," one British delegate was forced to plead.) The moment one defines neutrality, it loses its neutral nature and takes on the national or linguistic character of the definer. Commander Sampson, no mean debater, observed: "Since France today proposed neutrality, we may conclude that they have the necessary delegated power to fully consider and determine the main question before us—the selection of a prime meridian."

The American delegates, Abbe and Rutherfurd, asserted that neutrality, as defined by the French, was a fantasy. Every longitudinal meridian touches land at some point in its arc, and is thus rendered suspect. And if the French wanted a Pacific prime, it would still have to take its bearings from some land-based observatory, which would then render it the astronomical property of some nation. The meridian favored by France made its landfall on the Kamchatka Peninsula, just west of the Bering Strait and thus would be controlled by a Russian or American observatory. Cleveland Abbe demanded, "How long will it remain so [i.e., neutral]? Who knows when Russia will step over and reconquer the country on this side of Bering's Strait? Who knows when America will step over and purchase half of Siberia?" And he went on, Cleveland Abbe of the looping, ephemeral weather maps, Abbe of precision and mutability, Abbe the friend of black empowerment, the visionary: "Something must be found which is fixed, either within the sphere of the earth or in the stars above the earth." (Global positioning might, in fact, eventually eliminate the need for an earthbound prime meridian.)

The French, now down to their fallback position, suggested that if the Anglo-Saxon countries would adopt the "neutral" metric system, France might accept an English meridian. No, Abbe repeated, even the *metre,* that one-ten millionth of the quarter-arc of the earth as measured by the French in the last century, was a *French* meter. (The Germans had recently remeasured the arc and come up with a slightly different value; thus a German meter.) An American measurement would doubtless create an American meter. All measurement is deneutralized by properties of the measurer. In any event, Admiral Rodgers ruled speculations over the meter out of order.

Failing to convince the French of the logical impossibility of their stand ("I have listened to my learned colleagues," J. C. Adams recoiled, "and it turned almost entirely on sentimental considerations"), the combined forces of the United States and Britain then shifted the argument to terms more conducive to their own pragmatic natures. It was Commander Sampson who introduced the word that had the sanction of the President and the Secretary of State: "convenience." Neutrality may or may not exist, but practicality does, and for the aggregate convenience of the world, only one meridian satisfies that demand. This was the course of debate that lit the fuse.

Janssen retorted: "We consider that a reform which consists in giving to a geographical question one of the worst solutions possible, simply on the ground of practical convenience; that is to say, the advantage to yourselves and those you represent, of having nothing to change, either in your maps, customs, or traditions—such a solution, I say, can have no future before it, and we refuse to take part in it."

After three hours of give and take, and with the original motion on Greenwich still unresolved, as well as France's motion on absolute neutrality, the third session ended and a week's recess was called. Of all the sessions leading up to a vote, the third was the least conclusive but the most revealing. The French position was exposed as fundamentally self-interested, and its proponents no better at defining neutrality than the British and Americans had been in attacking it. And there was genuine passion, a kind of requiem, in the French defense of what they felt to be the surrender of their noble astronomical tradition on an altar of complacent Anglo-Saxon commerce.

When the delegates reassembled a week later on October 14, the weather had turned cool and misty, all traces of summer gone. Instructions from governments had been cabled, speeches rehearsed, and the swords were out. It was Sandford Fleming, France's erstwhile ally, who opened the argument. Pressing hard on France's greatest sensitivity, he displayed the figures of maritime tonnage that were then employing Greenwich charts. Paris came in second, one-fourteenth the size of Greenwich. Seventy-two percent of commercial shipping used Greenwich, 8 percent used Paris. But Fleming's purpose was not to attack Paris; rather, it was to eliminate a potential objection to his own anti-prime, which he then posed as the ideal compromise between utility (Greenwich) and utopia (neutrality). As always, he wanted the advantages of Greenwich, without calling it by the dread name.

His intervention was not warmly received. The French chose to see it as coming close to a parody of the classic split between two philosophical traditions: the sleek, rationalist vessel of polished *discours* rammed by the rusty coal barge of British empiricism. Finally, out of frustration, anger, and self-pity, the French gave up and called for a vote on the neutral meridian, which was defeated, 21–3. Fleming's compromise, had the French accepted it, was logical, even virtuous. The anti-prime retains all the charts of Greenwich, but removes from them the taint of a British connection. But the French refused to accept the argument; they saw the anti-prime as still infected by the English contaminant. The atmosphere had grown poisonous, and the British-American axis had been encouraged by the final vote, to adopt a triumphalist pose. Rutherfurd immediately renewed his Greenwich motion, Fleming amended it to leave room for his own anti-prime—to call it the "great circle" of Greenwich—but the British delegates, Evans, Strachey, and Adams, revolted against their Canadian member, announcing that they renounced Fleming's position. Fleming's amendment lost, and the rest is part of our common history. Even his later time-zone recommendations were ruled out of order. The French, and Fleming, lost.

And while he was at it, Rutherfurd exposed one rather obvious flaw in Fleming's carefully worked-out proposal. He pointed out that a Pacific prime would have disastrous effects on London, and Western Europe in general. If the anti-prime became the prime, Greenwich would become the anti-prime. At noon every day, England would be split into separate days. Clearly it was preferable to change the world date in the unpopulated Pacific than in the middle of the largest city in the world.

Greenwich passed. The world day, the counting of time zones, the astronomer's day, all civil time for all societies begins at midnight on the Greenwich meridian. Fleming's anti-prime disappeared, only to be resurrected at a later date as the international date line. It was left to the Russian ambassador (and amateur astronomer), Charles de Struve, to clean up computational anomalies left in the wake of the Greenwich selection. First of all, he proposed the rescinding of the Rome accord on the universal day (one of the last Fleming initiatives still standing). The universal day should begin at the Greenwich midnight, not noon, for the avoidance of double-counting the dates in Europe and North America. The Russian deserves credit for the east-west division of longitudes, 180 degrees east and west from the zero meridian, also an overturning of a Fleming-supported Rome agreement. Otherwise, Greenwich would have been in the anomalous position of lying, simultaneously, on the zero *and* the 360th degree of longitude. Villages a few miles west of Greenwich would carry unwieldy coordinates like 359 degrees west longitude, while neighboring towns a few miles east might lie at two degrees.

To de Struve also goes the credit for unifying the astronomer's day with the civil day ("We think it easier for the astronomers to change the starting point, and to make allowance for the twelve hours of difference in their calculations, than it would be for the public and for the business men, if the date for the universal time began at noon, and not at midnight") and, perhaps most significantly, for the creation of the all-important international date line ("the change of the day of the week, historically established on or about the anti-meridian of Greenwich, should henceforth take place exactly on that meridian"). He supported the adoption of the universal day (clause five of the Rome Conference), "side by side with the local time, for international telegraphic correspondence, and for through international lines by railroads and steamers." And he warmly supported Fleming's beloved twenty-four-hour clock. And so Sandford Fleming's great adventure had come full circle.

And why, I ask myself, did Fleming, and the French, lose so profoundly? In Rome, an understanding had been struck that England and America would give serious thought to adopting the metric system (it is still the legal standard in both countries, and was nearly enacted in the United States in the last century) in return for French acquiescence to Greenwich. That hope, along with the support shown in Rome for the anti-prime, certainly motivated French participation in the Washington Conference.

But why did Fleming fail to see the flaws in his position?

Blinded by vanity and ambition? Overmatched, finally, by the world's scientific and diplomatic elites? I think not. The French (and Fleming) were correct to insist on a "scientifically" derived prime. Science says only that all meridians are equal, and if the maritime tonnage alone is to settle the issue, then the conference need not to have been called in the first place. The French (and Fleming) were also correct to downplay the need for a first-class observatory on that prime, and correct to show faith in the accuracy of modern telegraphy. But the French were even more committed to scientific neutrality than Fleming; he, after all, was a secret partisan of preserving Greenwich charts for his anti-prime proposal.

It was politics that had given momentum to the concept of convenience. The indelicate French response, that the proud traditions of ten nations were to be sacrificed to the size of another meridian's clientele, points out a serious flaw. If convenience and popularity were the only standards that matter, then science had been replaced by political power, and by commercial mandate. For many reasons, the Prime Meridian Conference can be seen as one of the first "modern" moments in science.

The fact is, the Washington conference was a diplomatic meeting, and Fleming had no country, no support, no constituency, and no rapport with his fellow British delegates, who were dedicated, above all, to the selection of the Greenwich meridian. There had never been a single date-line meridian as there had been various primes; the Portuguese had marked Macao as their date line, the Spanish chose Manila, the Dutch, Batavia. Just as prime meridians had proliferated over the centuries, so had date lines. The exact link between the prime and its precise half-global anti-prime—the date line—had not been set until Charles de Struve articulated it. Fleming might be forgiven for concentrating on his Pacific prime, and neglecting the implications for a precise European date change. We might even see a time, in the not-distant future, when the change-of-date between North America and Asia is considered far more disruptive of commerce than would be a similar a separation from Europe. Splitting London (or even continental Europe) into different calendar dates is obviously unworkable, but Fleming and his allies, including such prominent geographers as Frederick Barnard, were committed to the system of "cosmic" time, a "world-day." Fleming in particular saw time as a continuous flow. Date-lines were an impediment, and their exact location could be negotiated. Ferro Island, in the Azores, had at one time served as the prime meridian; perhaps it could serve again as a date-line.

Naval astronomers acting on instructions from their national governments took the initiative in Washington. In Rome twelve of the thirty-eight delegates had been academics, directors of their national observatories, thus a little more independent of diplomatic concerns. In many ways the naval astronomers were the perfect technocrats for the job. They were informed, practical, and aggressive. They saw no virtue in the appeasement of sentimentality, they pressed their advantages and were indifferent to the diplomatic "susceptibilities." And that, I feel, was the major miscalculation of Sandford Fleming—his fear of arousing those sensibilities, and overestimating, perhaps, the persuasive (or disruptive) powers of the French, and the effect they might have on the necessary unanimity that standard time for the world required.

Men like Fleming, who had been so influential at Rome and Venice, had no role to play, except as foils, in Washington. Fleming ended up patronized by the British, ignored by the Americans, and held in contempt by the French. William F. Allen, a fellow engineer from the world of railroads, the man who had transformed time in North America and been hailed a scientific genius in the American press just six months earlier, spoke up only once during the entire three-week conference, to no practical effect. The major result of the Washington Conference, the fixing of Greenwich, was not scientifically or diplomatically settled. It was a popular vote. The arbitrary conditions of nine-teenth-century commerce have, however, defined time for over a century. Fleming returned to Ottawa without a comment in his journal, except to note that he had seven dollars in his pocket.

NOW, AT THE TURN of the millennium, Abbe's arguments on the impossibility of separating a definition from the definer, or neutrality from self-interest, objectivity from cultural bias, seem to evoke the unsteady first steps of that demon-in-waiting, the uncertainty principle. And in Fleming's appeal for the uni-versal day, cosmic time, and the anti-prime—"as speed increases, time appears to shrink, and space to expand"—the modern ear detects the distant rustle of relativity. The last year in human his-tory with too many times, but no place for starting and ending them, was over. A great hinge had creaked shut and the world had been fundamentally altered. Sundial time was banished, a sophisticated abstraction had taken its place. Not a penny was spent, not a drop of blood spilled.

And who were the winners in Washington? At first glance, Britain and the Anglo-Saxon world, of course. But did France, or Fleming, retreat in tatters? Hardly. History has enshrined Fleming and rendered the names of de Struve, Rutherfurd, Adams, Samp-son and the rest to honorable obscurity. He went on to enjoy the undeniable and wholly deserved credit for the trans-Pacific cable.

France abstained from voting throughout and true to Lefaivre's promise, "Greenwich" never appeared on any French chart. In 1898 official French time was defined as "Paris mean time, re-tarded by nine minutes, twenty-one seconds." Oddly enough, the "retarded" French time was thus identical to that of a certain leafy London suburb. But obstinacy has its triumphs, and com-placency its defeats. Janssen was correct. The electric telegraph could indeed have replaced a full-blown modern observatory; some outpost on Diomede Island in the Bering Strait could have served as the prime. France had taken the lead in scientific telegraphy and is today the home of Universal Coordinated Time (UTC), which has replaced GMT almost everywhere. The difference is less than a second per day; UTC enfolds the leap-second phenomenon within it, to compensate for the earth's decelerating rotation. The only country still using Greenwich time is Britain.

A HUNDRED and fifteen years after the agreements on standard time, familiar conditions are reappearing. I've called them "a-temporal" manifestations, but they're also commonly gathered under the rubric "the digital divide." To enumerate: cell phones, jet travel, e-mail, DNA analysis, computer-speed communication, storage, and retrieval—they all transcend the normal fixed-time, fixed-space concepts of the time-space continuum. Crimes committed decades ago and thousands of miles away can be solved. Phone calls can be made or received independent of a fixed location of sending or receiving. In Iowa, I have sat in the cab of a tractor and watched weather maps from Argentina, Australia, and the Ukraine flash across a computer screen, while updated commodity prices roll across the bottom, allowing an Iowa farmer to know if he should sell or withhold his harvest. I was on a farm (what could be more "natural"?), but also in the midst of a rational reconstruction of a farm, with the crops and equipment of a traditional farm, and men and women who looked and talked and lived like farmers, but who were also world citizens who visited their Chinese sister-cities in the winter and attended international conventions and, in the midst of their harvests, would host visiting foreign writers with lavish farm banquets. They were on first-name terms with their congressmen and senators in Washington, and testified at agricultural hearings with all the alacrity and pertinence of a Fleming or Abbe.

An influential world minority, "temporal millionnaires" live outside the rigid bars of standard time. For them, through jet travel and e-mail, time zones come and go like Abbe's dreamy isotherms. As with the nineteenth-century railroad grid, the Internet is clogged, but its partitioning, or any kind of regulation, is strenuously resisted. The work-at-home types, the twenty-four-hour stock traders, the freelancers, all live outside of their time zones. My San Francisco neighborhood is full of young men and women walking their dogs at eleven o'clock in the morning, enjoying brunch in sidewalk cafés. But don't be fooled: they put in their hours, some of them defined by the Tokyo markets, some by New York, many by an electronic tether to Silicon Valley, or even Bangalore. We have yet to devise a system of time that enhances the globalization we see around us wherever we go.

My darkened study is a winking harbor of red and green lights—nothing is ever off. The great toggle of the Industrial Age, on/off, has been overridden—the world is a vast network of sentinel technologies sleeping with one eye always open. To be "off" is to be unplugged, a way no one can afford to be. Airplanes communicate in Universal Coordinated Time, indifferent as the railroads had been to the time zones they're passing over. Someday, and I imagine it will not be distant, we will communicate directly in time, and through time, in some version of Fleming's universal day, a day whose coordinates are positioned overhead, satellite versions of Yeats's golden birds, meridians in the sky, out of nature all together (and do we still need the earth's rotation to establish a day?). Can't we recompose time itself?

We hear complaints of the "time bind," of running out of time. But time is what we make of it, and for a hundred years we have been stuffing the Industrial Age hour, the workday and workweek and the Industrial Age time zone with more and more work, more things to do, and faster, more efficient, ways of doing them. But the hour is not sacred. It reminds me a bit of the school day and the school year, which are being stretched now to accommodate more students and more learning. The time is out there, if we stretch it—or, once again, if we assert ourselves against its tyranny. Standard time, which had been the operating system of a new technology, is finally an obstruction to some, and an irrelevance to many more.

The dominant technology of our age, the computer, is restless. We're simultaneously running out of time, but no longer confined by it. The World Wide Web maintains its own "time-standards," an almost perfect echo of the railroad "time conventions" of the nineteenth century. We don't know what sleek new solution will come slouching out of Silicon Valley, but if it needs to change time again, it will. Fleming's cosmic time always had a Trekkie feel to it. Its defeat, in the very long run, might only be temporary.

Part Three

AFTER THE DECADE OF TIME

Britain, 1887

We were seated at breakfast one morning, my wife and I, when the maid brought in a telegram. It was from Sherlock Holmes and ran in this way: "Have you a couple of days to spare? Have just been wired for from the west of England in connection with Boscombe Valley tragedy. Shall be glad if you will come with me. Air and scenery perfect. Leave Paddington by the 11:15."

—from SIR ARTHUR CONAN DOYLE,
"The Boscombe Valley Mystery"

SHERLOCK HOLMES, that paragon of ordered time, was always sending telegrams to Dr. Watson. A plague of small boys was forever delivering them. There was a lot of to-ing and fro-ing by train from Paddington, Euston, or Charing Cross to deserted moors and country estates. So utterly reliable were trains by 1887 (the year of Holmes's literary debut), and so orderly the country-side, that on the way down to Brighton he could (in "Silver Blaze," 1890) entertain Watson by gauging minute fluctuations in their velocity merely by timing the telegraph poles, set pre-dictably at sixty-yard intervals, as they flashed by his window. A few quick calculations and he can announce, "Fifty-eight miles an hour," and a few minutes later, "Fifty-six." It is another metaphor for standard time itself: the train acting as the sun, the telegraph poles as longitudes. Time is the expression of velocity through measured space; velocity the division of time into distance.

It also tells a small inside story of time that will not surprise readers of this book, but might have eluded the master sleuth himself. The first use of bogies outside North America was on English Pullman cars, on the London-to-Brighton run. One reason for the accuracy of Holmes's calm calculation of velocity, and the confidence with which it is delivered, is the smooth-running, sturdy platform from which it was delivered. His off-the-cuff display of brilliance owes something to Ross Winans's invention of nearly sixty years before.

Conan Doyle not only created a memorable character in Sherlock Holmes, he was propagandizing for a modern, rational Englishman. The times in England called for a redemption of the national character. British self-confidence was fraying, a hero was needed. And not just any hero, but a perfectly rational, middle-class hero. Even now, there's something iconic about Holmes's glancing at a telegram and crumpling it up, or sending a response with its characteristic terseness and light touch ("Air and scenery perfect"). From it radiated a world of unflappable precision, withheld mystery, and repressed emotion that we associate with the perfect English gentleman.

There are only two things wrong with the portrait. By 1887 the bright sun of British scientific and industrial confidence, so engagingly captured in Holmes's certainties, was hanging very low and very dull in the western sky. The United States, Germany, and even a resurgent France had stolen Britain's preeminence in all but colonial holdings (which had become an increasing burden) and maritime tonnage. Holmes is a late-Victorian mock-up of vanished mid-Victorian confidence, accepted and beloved by a credulous readership anxious to indulge a new national myth.

He belongs with Stanley and Livingston, Sir Richard Burton and Gordon of Khartoum as a creature of the emerging media culture. And, second: so far as I can tell, Holmes was no gentleman. His class origins are mysterious.

It is not the presence of telegrams or railways that is striking—the technology of both were universal—but rather Holmes's icy confidence in their deployment. When he observes to Watson, "There is nothing more deceptive than an obvious fact," he was not doubting the 11:15 Paddington departure. He was attacking the complacent reliance on common sense, that lofty reluctance to engage the self-evident—in other words, the ancient enemy of rationality, the "natural." Its well-regarded American cousin, horse sense, led to faith in the *Old Farmer's Almanac,* and, even today, to "creation science" and advice from 900-numbers.

A Briton in 1887 could depend upon rail and telegraph networks not only because the world was unified under a single time, but also because Britain had been functioning under it nearly forty years longer than any other country. British legal, political, and commercial institutions, including Mr. Sherlock Holmes's forensic methods, had evolved with standard time, the expectation of temporal coherence, embedded within them. Holmes is a model of clear thinking and rationality—but so was British society. Anarchic, chaotic America was still adjusting to its railroad standardization of 1883 and, along with the better-organized Germany, to the prime meridian agreements of 1884.

PROBABLY NO character in literature has been hijacked more often by more critics and polemicists than Sherlock Holmes. (I'm doing it myself.) Holmes the cocaine-taking rebel, Holmes the outsider, Holmes the voice of reason, Holmes the cryptographer, Holmes the last, best chance against international and intergalactic terrorists—Holmes is cocooned from change. The cases he solved were clever for their time, but hardly the source of his interest today. He remains the essential Englishman—any

other nationality for him would be unthinkable—but the paradox he represents, minus some of his insufferable attitudes, has
been adapted repeatedly and woven into the subtext of any number of American genre films—detective, western, even contemporary space dramas. Any hero who doesn't get (or even *want*)
the girl has a bit of Holmes in him. So does the incorruptible private eye of film noir, and the brave sheriff, the Lone Ranger, the
computer, and the robot, even the android, blessed with every
mental and physical power known to science but willing to surrender it all, including immortality, just to become more human.
There's a contemporary feel to Holmes; we can imagine him
belonging here and now. New adventures are constantly being
written for him, featuring new and worthier opponents; new actors play him, and he creeps closer to us all the time, fighting
Nazis, fighting future villains. We want to keep him around, if
only to crack his reserve. Maybe *this* time he'll open up, find a
woman worthy of him, a challenge big enough to engage his full
humanity.

When I wrote earlier of time and aesthetics, I'd had no ready
category for the quality Sherlock Holmes represents; he's a
category unto himself. Certainly he embodies the late-Victorian
Zeitgeist; certainly his character suggests something of oracular
timelessness. His salient quality is that of *non*-representation,
he rides above time, even if a plethora of period details speak to
the contrary. The truth is, Sherlock Holmes is literally timeless,
without a time of his own, still out there, looking. More than a
century after his first appearance, his appeal is undiminished and
his attitudes seem to apply to more causes and to uphold more
positions today than many of our contemporaries, let alone a figure from history. Is there another character from nineteenth- or
early-twentieth-century literature with quite his vigor, quite his
chameleon nature? Bartleby? Kurtz? Gregor Samsa? Leopold
Bloom?

And yet—here's the paradox—has there ever been a more unpleasant, a less attractive figure than Sherlock Holmes? Of all the heroes in all the literatures of the world, none quite has his singular lack of appeal. He is utterly lacking in humor, in charm, in social graces, and sexuality. He lacks a gentleman's modesty, and apparently has suffered no inner struggle to achieve the particular pinnacle he occupies. Even his creator came to despise him.

THE PSYCHIC rumblings of locomotion hardly affected Mr. Sherlock Holmes. Trains and telegrams existed simply to shorten the hours or days between the commission of a crime and the commencement of an investigation. Crimes could only be solved rationally by reversing time, returning the crime scene to the moment of murder, the killer still present, the knife raised, the revolver pointing.

What is a clue, after all, but something out of place, or out of time, present where not expected, or absent when anticipated? Waste a moment, or let the doltish Inspector Lestrade (another holdover from the natural world) tramp over the evidence, let it rain, and the time-space continuum is broken, or even worse, is re-formed around the anomaly, and the clue (as clue) disappears. Holmes was in a hurry to get to the scene, even checking the evening trains in order to sneak in an interview with the prisoner, and still return for a late dinner. "I say, Holmes, what's the rush?" asked his faithful companion. *Indeed,* Holmes might well have answered, *a change in the pace of change, my dear Watson.*

Reading Holmes after so many years, I am struck not so much by the free application of so much science, so much rationality, and so much agnosticism—those prime Victorian virtues—but by how far they are from the natural-world assumptions of an earlier practitioner of the same trade, "the father of the murder mystery," Edgar Allan Poe. They write of the same mysteries, but from opposite sides of the standard-time divide. Poe's investigative assumptions are more atmospheric than logical, procedural, or sequential. Criminals are, in noticeable ways, exogenous to humanity (even at times an ape), not, as Holmes assumed, hospitable tea drinkers just like us. In Poe, we are struck by the murderer's freakishness, the pallor and nervousness, the trembling fingers. In terms of this book, it could be said that Poe's "rational" consciousness, like Bartleby's, was struggling to cope with the imaginative limits of a "natural" world.

Poe did produce one tale (regrettably, pale and didactic), "Three Sundays in a Week," that directly discusses temporal confusion in a prestandardized world. Without an international date line, one circumnavigator of the world leaving London in an easterly direction "gains" a day in the completion of his journey. Another, going west, "loses" a day. When they return to London on a (London) Sunday, one sailor's Monday registers in London as a Sunday, as does the other sailor's Saturday. Thus, Poe earns his title, and one intrepid sailor wins a bet and gains the hand of a desirable woman. Three separate days are all Sunday or, conversely, one day is recorded as three, and all of it happened thirty years before Professor Dowd's experience with the three separate clocks in the Buffalo train station.

A minor quirk of history reinforces the same point. When the United States purchased Alaska in prestandardized 1867, the Russian Orthodox inhabitants of that far eastern province found themselves suddenly having to observe the sabbath on the American Sunday, which was Monday by Moscow reckoning. In the end they were forced to petition the patriarch for guidance as to when to celebrate mass—the Russian Monday or the American Saturday.

HOLMES'S CERTAINTIES ("The murderer is a tall man, left-handed, limps with the right leg, wears thick-soled shooting-boots and a grey cloak, smokes Indian cigars, uses a cigar-holder, and carries a blunt pen-knife in his pocket") are almost patriotic outbursts, as reassuring to his readers as *Idylls of the King* or *Bell, Book and Candle.* Conan Doyle was well aware of his society's predilection to dream of vanished glories and to trust in celebrity saviors rather than confront social problems and undergo necessary reforms.

He also understood that Victorian rationality had been purchased at the price of suppressing its natural, darker urges. Accordingly, he created Professor Moriarty, a Victorian Darth Vader, the embittered, excluded twin of progress and enlightenment. Reason and progress had created a pool, of nearly equal depth and expanse, of rejection and reaction. Conan Doyle also understood a commercial truth: a super-sleuth must finally tangle with, or even perversely create, an opponent fully worthy of himself. The Victorian public, however, did not appreciate Doyle's literary balancing act, nor did they accept the yin and yang of Victorian popular culture obliterating themselves at the Reichenbach Falls. To his author's great distress, Holmes was brought back by popular demand; Conan Doyle extracted his revenge by retiring him, eventually, to a life of perfectly sane, perfectly rational, perfectly predictable, beekeeping. In bees, and bee culture, he presumably found his reflection in the world.

Holmes and Moriarty's mutual cancellation would be a closer rendering of the historical record. If Freud and Einstein, the ultimate rationalists, represent the spirit of Sherlock Holmes, then Moriarty is a creature of the clogged drain, that backup in the basement of reason. He represents the backside of science that would give us the twentieth century's endless supply of monsters and the ideologies that back them up. Though devilishly clever (and himself the subject of innumerable updates), he is the twisted twin of agnostic rationality. Freud's *The Interpretation of Dreams, The Fundamentals,* and Einstein's Special Theory of Relativity all appeared between 1900 and 1905. A hundred years later—and this would have astounded Victorian thinkers like Charles Kingsley, or the grand old apostle of science, Thomas H. Huxley—politicians, the media, and a divided citizenry are having to make greater accommodation for the fundamentalist agenda than for the work of Darwin and Victorian humanism. Faith in progress had suppressed only temporarily the progress of faith.

NEARLY HALF a century after the arrival of standard time in Britain, at Victoria's Diamond Jubilee in 1897, England's supremacy, her wealth and confidence, and certainly the extent of her empire still seemed unchallenged. Her cultural and scientific achievements during those five decades are among the greatest ever recorded.

In the shadows of her monuments, however, stirred the subversive architects of Britain's coming decline. The ancient, unaddressed evils of the class system collided with the modern imperatives of technological investment and educational opportunity. Never having experienced the hand of destruction, or even significant challenge, England's institutions were still intact. She had not been forced to confront the extent of her unpreparedness. Predictions of collapse, from the likes of H. G. Wells, seemed excessive.

In the early part of the nineteenth century, Britain had invented most of the world's technologies, primarily the steam locomotive and the telegraph, along with coal-based dyes and new forms of metal- and glass-working ovens. By the 1880s, however, she was importing German chemicals (the successors to dyes) and delicately calibrated machine tools from America. Britain lost control of the industries and the technologies that she'd created, and along with them, she lost control of her future. The "civilizing mission" of imperialism, so readily subscribed to earlier in the century, so inspiring to thousands of idealistic best and brightest colonial administrator-missionaries, had become a cynical exercise in trade and despotism, a holding action against rapacious exploiters like France, Belgium, Portugal, and Germany. Britain had been exporting her most energetic young men and women to the colonies and the United States (largely her marginalized Scots, Irish, and Welsh) for over a century, retaining a cadre of impeccably trained gentlemen to run its home offices, industries, and colonial outposts. The results were proving disastrous to Britain, and often to her colonies.

No wonder Holmes's audience clung to his every word. He reembodied Britain's self-image of no-nonsense authority. He was never kindly (that was the role of the good doctor), but he was always fair. So long as he represented his country's rational supremacy to the world, society was spared from having to make unpopular and uncomfortable reforms. Even if the implications of his methods are authoritarian in the extreme, Sherlock Holmes is the touchstone of all that was good and decent in that quality universally recognized as English. (And when most needed, Dr. Watson is in attendance to supply a quick dose of humanity.)

"The Boscombe Valley Mystery" turns on a seemingly open-and-shut case of parricide in the west of England. A sure sign of the misreading of evidence is Lestrade's assurance of the prisoner's guilt. "Open-and-shut" refers more to the mind of the investigators than the strength of the case against their suspect. A young man will go to the gallows for his father's murder unless exculpatory evidence can be unearthed (literally) by the celebrated London sleuth.

Holmes is urbanity personified, alien to the outdoors and slightly distressed by it. It's that distance that allows him to read nature rationally. Watson records: "He drew out a lens and lay down upon his waterproof to have a better view, talking all the time rather to himself than to us." Victorian forensics might not have had access to fingerprints and DNA, but it did have railways and telegraphs, and magnifying lenses to pick up fibers and cigar ash and traces of mud and bits of extraneous material whose presence (or absence) could speak volumes to a mind prepared to hear their stories. Holmes is the master of drawing out significant presence from apparently absent, or irrelevant, facts. And in the case of the Boscombe Valley murder, the unearthed evidence proves that there had been a lurking presence behind a tree before the son's appearance, and that in times past the victim and the murderer had both lived, and feuded, in Australia. The simplest shred of evidence eventually unravels the world, and is that not the enduring appeal of Sherlock Holmes?

A SMALL CASE from the annals of Dr. John Watson, recounting his years with Holmes. Order and reason prevail, under the authority of *a posteriori* logic and inductive reasoning. Should a white shark have appeared on an English country road, Holmes would not have leapt to a Fleming-like conclusion. Or closer to the point, if a strange-looking skull had turned up in a local bog, near a village called Piltdown, he would have guarded his enthusiasm. He would stretch out in the mud, take out his glass, and declare that nothing is more deceptive than an obvious fact. He would have declared it a fraud, not the missing link.

Time, Morals, and Locomotion, 1889

Time travels differently when you're on a train!
—Advertisement for Amtrak, 1999

WE THINK OF the closing of the frontier as a North American phenomenon, but Europe had an eastern frontier no less formidable than anything in the American West. In 1883, with "time in the air" and American railroads adopting standardization on the Sunday of Two Noons, selected members of the European press, mainly the society columnists, were assembled for the jaunt of a lifetime, the first run of the Paris-to-Constantinople Orient Express. (Regular service for civilians would not start for another five years.) The Balkan region was a minefield of outlaw activity, but the luxury train skittered across it as blandly as a luxury liner through a field of icebergs. The sheer audacity of the enterprise, bringing "the Orient" in touch with Paris, was a dream fulfilled, a link that had eluded Europeans, and their invaders, since the fall of the Roman Empire.

As we already know, however, European railway history did not favor the larger, slower, American cars. But suddenly, a part of Europe opened up that was strangely like North America: the Balkan region was extensive, rugged, and relatively underpopulated. Land was cheap, so there was no need to engineer the shortest and straightest route, which, given the mountainous terrain and its recalcitrant inhabitants, would have been prohibitively expensive. When a pair of return tickets between Paris and Istanbul cost the equivalent of a six months' lease on a Mayfair house, and many of the passengers were bankers, arms dealers, casino habitués, and their consorts, so long as the champagne and caviar held out, what was the rush? American and European styles of speed and luxury, separate for fifty years, had merged. Europe now had American design and technology. They had adapted it, however, to a class system at perhaps its height of decay.

The Orient Express banished all notions of time and space. The languages of its staff changed with each new border crossing. Turkish effendi roamed the corridors, adept in every language, addressing the desires of every treasured passenger. The result of luxury wedded to speed on both continents was a new form of time reversal. Outside, it might be the wild prairie a thousand miles from civilization, or a Turkish-dominated Europe seething with rebellion, but inside it was a Chicago saloon or a five-star Paris hotel. After 1881 in America, Pullman cars boasted electric lights (still a rarity in most homes), cold champagne, and WCs with flushing toilets. After 1887 on both continents, the cars were totally electrified. The air was seasonally adjusted and kept circulating. Travelers could move across the continents under the same, or superior, conditions than those at home, bringing only the more benign forms of nature indoors. Just about the only thing that had not been interiorized was time itself.

THE VANDERBILT of southeastern Europe, the major financier of eastward railway expansion, was one of those fabulous visionaries of the nineteenth century who casts a long shadow into the present era, the banker and philanthropist Baron Maurice de

Hirsch (or Moritz von Hirsch). One of de Hirsch's undertakings of lasting significance to history was his scheme for Jewish resettlement, bringing entire shtetls from Russia and the Turkish provinces to the recently opened new lands in Saskatchewan, Argentina, and Brazil (opened as a result of universal railroad expansion). In bringing Western Europe to the East, and relieving intolerable conditions in the East with the freedom of the west, de Hirsch weighed profit against philanthropy much in the way of other mighty barons of industry, the Fords, Rockefellers, and Carnegies.

The baron was driven by a vision of continental unity even before the Russian and Turkish lands were thought of as being connected to Europe at all. He realized that railway expansion had to play the central role in any kind of unification. The negotiations between Turkey and Russia, France and England, Germany and Austria—as well as the attendant dangers posed by outlaws and revolutionaries across that tortured region—stretched on for twenty years, pushed and prodded by de Hirsch, with long recesses for wars and treaty negotiations.

The fifty-year process of moral rationalization from 1860 to 1910, the often flawed attempts to right some of the "natural" wrongs of history, as in the antislavery battle in the United States, the lifting of anti-Semitic restrictions in Central Europe, the national unifications of Germany and Italy, the breakup of Austria's and Turkey's Central European and Balkan provinces, are some of the political accompaniments to the long process of temporal standardization. Speed burst the imperfect welds of history. Unstable identities, as Schama has noted, were preyed upon by history. Except on the African continent and in parts of Asia, the rights of minority populations to express their cultural identities and to assert their political wills was at last recognized.

For the first time in two thousand years, thanks largely to a visionary banker who hailed from the fringes of accepted European society, and to an American industrialist whose labor practices have earned him permanent disgrace (Pullman), a dream that had survived since the Romans, Alexander the Great, and the Holy Roman Empire had finally come to pass. By 1891, *wagons-lits* cars on regular trains, some on the baron's lines, were available all the way from Lisbon, Madrid, and London to Moscow and St. Petersburg, by way of Paris and Vienna. Berlin remained suspicious of Russian penetration of its territory and did not participate until 1896. In 1898 luxury service was available as far east as Tomsk, in mid-Siberia. Railways had gradually unified the continent. Europe, ever so briefly, and ever so perilously, was one.

In sixty years, a generous human lifespan in the nineteenth century, locomotive technology evolved from Stephenson's "Rocket" (1828), on which a passenger might be treated as tenderly as the lump of coal that propelled him, to the super-trains of the Orient Express. The fully evolved *wagons-lits* resembled nothing so much as Fleming's 1863 vision of "floating hotels crossing the Atlantic," like evenly spaced nodules pulled along a single track.

THERE IS FAR more to the railroad revolution than technology or diplomacy. There is the question of morals—particularly, of sexuality. The combination of speed and luxury, with its resulting mobile society, inevitably calls to question the traditional proprieties. The mountebanks and reprobates on the Orient Express were legendary even in their own time; we'd recognize them today—they are not the story. The real story lies in the making of a new morality. Think of a short-haul, mid-American day train. No Pullman luxury. No one rich and famous, just those sturdy American archetypes, the traveling salesman and a farmer's daughter.

In August 1889 a "bright, timid" eighteen-year-old, small-town Wisconsin girl by the name of Caroline Meeber kissed her family goodbye, shed a tear, and boarded the afternoon train for

Chicago, where she intended to live with her married sister while seeking work in the city. It is one of the oldest American stories, one of endless *becoming*, leaving the closed-in town for the city, seeing a bit of life, finding a job, and probably a husband, where the opportunities were broader. Most of those stories, however, start (and often end) in that stifling small town, or in the dark and dangerous city. Very few pick up on the transition zone between town and city, the way Theodore Dreiser did in *Sister Carrie*, published in 1900.

Even before reaching Chicago, Carrie meets a glib-tongued traveling salesman, a "drummer," by the name of Charlie Drouet. He gains her trust (trust being the only thing she has to give, having trusted everyone for eighteen years), and wheedles her sister's Chicago address. Carrie works honorably for a few weeks as a seamstress, her eyes straining in the poor light, her back and legs aching. Her sister's husband cleans cattle cars down at the stockyards, with predictable effects on his disposition and domestic behavior. She learns quickly enough that there is no honor in honest labor, and that women so employed yearn to be delivered from a life of brutal exploitation, even at the loss of their virtue. When Drouet reenters her life, she's seen enough of her sister's condition and is ready to leave. She becomes a kept woman, passes from man to man, rising each time ever higher on the social ladder, and eventually finding the niche that nineteenth-century society provided for canny and attractive young women. Her story was a scandal that had to be censored in its day.

By their lights, authorities were right to ban it, and the publisher did the decent thing by withdrawing it. Dreiser had taken a familiar backstreet story, easily dismissed as vulgar and distasteful, added a new ingredient, and made it relevant to every living room in the country. The ingredient he had added was *speed*. Just as Sherlock Holmes had identified possible suspects as "one of us," so had Dreiser created a fallen young woman who looked, sounded, and acted precisely like us, a healthy and confident girl from the edenic heartland, with supportive friends and family. She was no tubercular wraith from the slums of the city, as Stephen Crane had written in *Maggie,* her appetites had not been corrupted by bad genes, poverty, alcohol, or abuse. Carrie's sin was knowing what she wanted, and what she had to bargain with, and how quickly she acted upon it. Her fall and subsequent rise, the surrender of virtue after the injustice of underpaid labor, came with a passive rapidity that was shocking.

Carrie is irresistible, not in the way of an attractive young woman, but irresistible as a force, like a locomotive at full throttle. Dreiser's conviction that female sexuality is no different from male sexuality was an idea whose time had not yet come in America. But sex is only the lure. Where the critics have underserved him is in emphasizing Carrie's sexuality, not Dreiser's radical analysis of social instability that had come about as a result of speed, a change in the pace of change.

The avant-garde doesn't always look shockingly new. Sometimes it lumbers around in earnest, sober, institutional prose. The new century in America was greeted by a revolutionary work that looked like, and sounded like (its critics charged), a lame, Midwestern imitation of Zola, or Thomas Hardy, slightly less didactic than Frank Norris or Upton Sinclair, nowhere as lyrical as Jack London or Stephen Crane.

In Dreiser's naturalistic universe, two moral codes (like two velocities) cannot coexist. The stronger, however one defines it—the cruder, the hungrier, the more sexually satisfying or more life-affirming, or, in terms of this book, the more energetic, the faster—must always triumph. Much later in his career, in *An American Tragedy,* he opened on an even more explicit image of the same conflict: on a cold city street, a family of evangelicals peddle their piety in music and pamphlets, posing a moral challenge to indifferent urban values. One of those child-evangelists grows up to murder his pregnant girlfriend. *It's all about time,* about the clash between rationality and the natural world.

THEY'D ALWAYS been out there in dirty jokes, but it had taken a train to bring the traveling salesman and the farmer's daughter together in a serious novel. For Drouet, train time was frame time, part of a performance. His whole existence was defined on the move, in self-presentation. For Carrie, new perceptions of reality altered old perceptions of self. She was a different person the moment she stepped aboard, her upbringing now irrelevant, and the brimstone certainty of retribution as well. Even an eighteen-year-old farm girl could buy a train ticket to the nearest city, labor a few weeks, reach a decision (however instinctive) about personal behavior and conventional morality, and set out on a life of endless self-discovery.

What Dreiser perceived in 1900 was a fact of life that society did not (and American society still does not) want to face. It is the jolt from the friction and collisions of the daily energy flux, the speedup between cause and effect, the expectation of instant gratification and the technology to deliver it that brings on panic and social change. The ingredient that Dreiser had added to the novel was moral velocity, and a character with the instinctive ability to understand it and profit from it. It is ironic, but predictable, that a culture like America's, so devoted to innovation, so proud of its impatience, so easily bored, is horrified to find that its core values (the remnants of the "natural" world) are continually under assault. *You can have speed, or you can have tradition, but you can't have both.* Or, as Werner Heisenberg phrased it in the uncertainty principle, you can know position, or velocity, but you cannot know both.

For some, like Carrie, speed defined the new authority and undermined old inhibitions. To keep on top of events when the events themselves are whirling faster than the human mind can comprehend required more work, more effort, more time, and less attention to tradition, or even to family, than ever before. What can tradition teach us, when everything is new? What respect is owed to outmoded thinking? The old ways of behaving, the proprieties that built the country, no longer applied. It's easy to salute, in a business-school model, the energetic few who broke the bonds of class structure and rose to wealth and power by shrewd instinct and ruthless self-discipline, but we're likely to underestimate the phenomenal balance it required, keeping atop the waves that swamped so many others. To be formed in such an era, and to survive it, even prosper under it, required a protean, assimilative nature, like Carrie's.

It's not that Carrie is nobler or coarser, or more or less intelligent, compassionate, generous, or inhibited than anyone else. She did not desert her parents or her sister; she simply moved away from them, at greater velocity. In her bland, unaffected way, she had mastered the change in the rate of change. The shocking thing about *Sister Carrie* is that our little sister started out more innocent and backward than any of the men who thought they possessed her, but somehow catapulted above and beyond them. They remained baffled by the changes in her, and in the collapse of their own fortunes. After all, they had given Carrie her start, they'd seen her first, and had laid their bets down on her. They felt somehow betrayed. And they didn't even realize that she had not changed at all.

What Carrie discovered in herself is the worst news that middle-class American society could have imagined. In Dreiser's words, she was a "pleasure-seeker." Innocent eighteen-year-old farm girls from Wisconsin could be pleasure-seekers. By comparison, the men in her life, especially those from the consumer class, like Hurstwood, a saloon manager who gives up everything to possess her, are "comfort-seekers." Society, as Freud sketched it, muffled the unruly libido, the pleasure principle, while elevating the reasonable, marketable, self-protective ego, precisely in order to protect its Hurstwoods from its Carries. But speed upset the balance, brought the pleasure-taking and pleasure-giving predators out of the shadows into contact with the placid herds of polite society.

Speed eroded traditional morality. It's hard to follow somebody who lags behind, whose values no longer apply. In Dreiser's naturalistic universe, ruled by the swifter, the more powerful, the keener appetites, pleasure trumps comfort every time. That same awareness had been dramatized half a century earlier in a number of near-contemporaneous works, such as *Madame Bovary* (1857) and *The Scarlet Letter* (1850), and in the work of a host of English and American naturalists, memorably in Hardy, and almost hysterically in D. H. Lawrence. Female sexuality is present as an unwelcome guest in the later Henry James, and *The Waste Land* is a recoil from it and its associated shredding of culture, ("O O O O that Shakespeherian Rag— / It's so elegant / So intelligent"). Eliot's proud conversion (or retreat) to classicism, royalism, and Anglo-Catholicism was a renunciation of any place in a "rational" world, a search for refuge in something resembling the sentimental shreds of the "natural."

Afterword

THE GHOST OF SANDFORD FLEMING

Time goes, you say? Ah, no, alas, time stays; we go!
—AUSTIN DOBSON on Lorado Taft's sculpture
The Fountain of Time, Chicago

HERE IS MY OWN eerie little time story.

In 1997, according to United Airlines, I circled the globe the equivalent of five times. I was fifty-seven years old, and director of the International Writing Program at the University of Iowa. Author recruitment, fund-raising, and literary festivals kept my bags packed and visas up to date. My wife, also a world-traveler, was a professor at Berkeley, two thousand miles away. She was born and raised in India, where we keep annual commitments. I was working on a second volume of autobiography. The first had concerned my father's dark, "natural" turn-of-the-century rural Quebec. One summer night, I was thinking about my mother's relatively sunny "rational" Manitoba childhood, her art-school years in England and pre-Nazi Germany, and her return to pre-war Montreal, where she'd met and married my father. A classic natural/rational matchup of contemporary worlds, I might say today.

My memory focused on a day in 1947, in central Florida, soon after our moving there. My father and I were standing on the art deco Main Street of Leesburg, next to our prewar Packard. He was dressed in his bright, Harry Truman–style Hawaiian shirt and high-waisted gabardines, stuffing pennies into a parking meter. I asked, "How can they be renting time?" And he'd answered, "They're renting space. It just comes out time."

A little later, the Ku Klux Klan staged its annual unmasked parade, leading to a baseball game. It must have been Confederate Memorial Day or Jeff Davis's birthday, one of those muscular displays of white supremacy the so-called New South lately tries to repress. That summer night in Iowa in 1997, as I watched a televised baseball game and read over the day's writing, two words, "time zones," started flashing on the page, as though a cursor had stopped in front of them and frozen. "Our lives are time zones," I'd written, "permitting the same things to be true and not-true, the same things to be here, and not-here." And I wondered, idly, why do those words suddenly seem strange, where did a term like "time zone" originate? The encyclopedia informed me that time zones were born with the Prime Meridian Conference of 1884, in which standard time for the world was decided. The leader of the movement was a fifty-seven-year-old(!) Canadian (bingo!) named Sandford Fleming.

Time zone seemed a brilliant portmanteau. Time doesn't have zones, I reasoned, but once we create them, all things are possible. Because 1997 was the fiftieth anniversary of Jackie Robinson's breaking of baseball's color bar, every baseball telecast that summer was full of Robinson footage, his steals of home, the clubhouse champagne, and the bright smiles of baseball's first black player. And every time I saw them, I thought back jealously to *my* Jackie, the times I'd watched him play in Pittsburgh in the fifties, when he'd torn up the basepaths, shredding the Pirates, and then back to my first baseball game in Montreal, in 1946, a year before his major-league debut, when Jackie had played for Brooklyn's top farm club, the Montreal Royals, and my father had taken me out to old Delormier Downs to see him. And there was a third Jack, this time in 1963, when we'd actually touched and said a word or two. Just after Martin Luther King's "I Have a Dream" speech, he was walking down the sidewalk, waving at well-wishers and shaking hands, a slow (Jackie, slow!), hunch-shouldered, white-haired Republican on the Democrats' big day. I told him I'd seen his *real* debut, and in that hard, tinny voice of his he said, "Montreal. Nice town. I enjoyed playing there."

Jack and me. I was, in one mystical moment that night in Iowa City, fifty-seven, six, seven, a teenager, and twenty-three. And now I was the only survivor. I abandoned the book I'd been writing, quit my job, and moved out to California to join my wife.

Bibliography

I HAVE PROFITED from readings in the Fleming holdings of the National Archives of Canada, and newspaper files from the Decade of Time. The works of Eviatar Zerubavel, for anyone interested in the pervasiveness of time (and the perversity of its measurements), are highly recommended. Fleming, a meticulous archivist himself, kept copies of the proceedings of various international conferences, including those of the American Metrological Association and, in particular, the Prime Meridian Conference of 1884. It was with great reluctance that I pulled myself away from those boxes of files at the close of each long archival day in the spring and summer of 1998 in Ottawa. I profited as well from the generosity of Kathleen Ryan Hall of Queen's University, Kingston, Ontario, for permission to study the notes of the late Professor Mario Creet.

Abbe, Truman. *Professor Abbe . . . and the Isobars: The Story of Cleveland Abbe, America's First Weatherman.* New York: Vantage Press, 1955.

Allen, William F. *Standard Time in North America, 1883–1903.* New York: American Railroad Association, 1904.

Altick, Richard D. *Victorian People and Ideas.* New York: W. W. Norton, 1973.

Amis, Martin. *Time's Arrow.* New York: Penguin Books, 1991.

Attali, Jacques. *Histoires du Temps.* Paris: Livre de Poche, 1982.

Barnett, Jo Ellen. *Time's Pendulum.* New York: Harcourt, Brace, 1998.

Basalla, George, ed. *Victorian Science: A Self-Portrait from the Presidential Addresses of the Presidents of the British Association for the Advancement of Science.* New York: Doubleday Anchor, 1970.

Behrend, George. *Luxury Trains from the Orient Express to the TGV.* Paris and New York: Vendome Press, 1977.

Berton, Pierre. *The National Dream.* Toronto: Penguin Books Canada, 1970.

Boorstin, Daniel J. *The Discoverers.* New York: Random House, 1983.

Brand, Stewart. *The Clock of the Long Now: Time and Responsibility, the Idea Behind the World's Slowest Computer.* New York: Basic Books, 1999.

Bruck, H. A., and M. T. Bruck. *The Peripatetic Astronomer: The Life of Charles Piazzi Smyth.* Bristol and Philadelphia: A. Hilger, 1988.

Burpee, Lawrence, J. *Empire Builder, the Life of Sir Sandford Fleming.* 1915.

Clifford, William Kingdom. *Lectures and Essays,* edited by Leslie Stephen. London: Macmillan, 1879.

Conrad, Joseph. *The Secret Agent.* 1905.

Corliss, Carlton J. *The Day of Two Noons.* Washington: Association of American Railroads, 1953.

Doyle, Arthur Conan. *Sherlock Holmes: The Complete Novels and Stories.* 2 vols.

Dreiser, Theodore. *Sister Carrie.* 1900.

Dyson, Freeman. *From Eros to Gaia.* New York: Pantheon Books, 1992.

———. *Disturbing the Universe.* New York: Basic Books, 1979.

Everdell, William E. *The First Moderns: Profiles in the Origins of Twentieth-Century Thought.* Chicago: University of Chicago Press, 1997.

Fabian, Johannes. *Time and the Other: How Anthropology Makes Its Object.* New York: Columbia University Press, 1983.

Faulkner, William. *The Sound and the Fury.* New York: 1929.

Ferris, Timothy. *The Whole Shebang: A State-of-the-Universe(s) Report.* New York: Simon & Schuster, 1997.

Fleming, Sandford. *Report,* Canadian Pacific Railway, Ottawa, 1876.

———. *Report,* Canadian Pacific Railway. Ottawa, 1877.

———. *Report,* Canadian Pacific Railway. Ottawa, 1888.

———. *The Intercolonial, A Historical Sketch.* Dawson Brothers, 1876.

———. *From Westminster to New Westminster.*

Fraser, J. T. *Of Time, Passion, and Knowledge: Reflections on the Strategy of Existence.* New York: George Braziller, 1975.

Gay, Peter. *The Bourgeois Experience: Victoria to Freud.* 5 vols. New York: W. W. Norton, 1984–1998.

Gell, Alfred. *The Anthropology of Time.* Oxford and Providence: Berg Publishers, 1992.

Gibbon, John Murray. *The Romantic History of the Canadian Pacific.* Tudor Publishing, 1937.

Gould, Stephen Jay. *Questioning the Millennium: A Rationalist's Guide to a Precisely Arbitrary Countdown.* New York: Harmony Books, 1997.

Grant, Rev. George M. *Ocean to Ocean, Sandford Fleming's Expedition Through Canada in 1872* (reprinted from the 1872 original in Cole's Canadiana Collection).

Gwyn, Sandra. *The Private Capital.* Toronto: HarperCollins, 1984.

Harris, Errol E. *The Reality of Time.* Albany: State University of New York Press, 1988.

Harvey, David. *The Condition of Postmodernity.* Malden, Mass.: Blackwell, 1990.

Hawking, Stephen. *A Brief History of Time.* New York: Bantam Books, 1988.

Hochschild, Adam. *King Leopold's Ghost.* Boston: Houghton-Mifflin, 1998.

Hood, Peter. *How Time Is Measured.* New York: Oxford University Press, 1969.

Horgan, John. *The End of Science.* New York: Broadway Books, 1997.

Houghton, Walter E. *The Victorian Frame of Mind 1830–1870.* New Haven: Yale University Press, 1957.

Howe, George F. *Chester A. Arthur: A Quarter-Century of Machine Politics.* New York: Dodd, Mead, 1934.

Howse, Derek. *Greenwich Time and the Longitude.* London: 1980 and 1997.

Huxley, Thomas Henry. *Collected Essays.* 4 vols. New York: Appleton, 1896.

Innis, Harold A. *A History of the Canadian Pacific Railway.* Toronto: University of Toronto Press, (reprinted) 1971.

James, Henry. *The American Scene.* 1904.

[Concerning Janssen, Jules-César]. *Homage.* Paris, 1922.

Kakar, Sudhir. *Frederick Taylor: A Study in Personality and Innovation.* Cambridge, Mass.: MIT Press, 1970.

Kaufman, Gerald Lynton. *The Book of Time.* Julian Messner, 1938.

Kern, Stephen. *The Culture of Time and Space 1880–1918.* Cambridge, Mass.: Harvard University Press, 1983.

Kubler, George. *The Shape of Time: Remarks on the History of Things.* New Haven: Yale University Press, 1962.

Landes, David S. *Revolution in Time: Clocks and the Making of the Modern World.* Cambridge, Mass.: Harvard University Press, 1983.

Lee, Samuel J. *Moses of the New World: The Work of Baron de Hirsch.* New York: Thomas Yoseloff, 1970.

Levine, Richard. *A Geography of Time.* New York: Basic Books, 1997.

Lightman, Alan. *Einstein's Dreams: a Novel.* New York: Pantheon Books, 1993.

Lightman, Bernard, ed. *Victorian Science in Context.* Chicago: University of Chicago Press, 1997.

Loizou, Andros. *The Reality of Time.* Gower, 1986.

Lomazzi, Brad S. *Railroad Timetables, Travel Brochures & Posters.* Spencertown, N.Y.: Golden Hill Press, 1995.

MacCormac, John. *Canada: America's Problem.* New York: Viking Press, 1940.

Marx, Leo. *The Machine in the Garden: Technology and the Pastoral Ideal in America.* New York: Oxford University Press, 1964.

Morris, Richard. *Time's Arrows: Scientific Attitudes Toward Time.* New York: Simon & Schuster, 1985.

Nelson, Daniel. *Frederick W. Taylor and the Rise of Scientific Management.* Madison: University of Wisconsin Press, 1980.

Newcomb, Simon, and Holden, Edward S. *Astronomy for High Schools and Colleges.* New York: Henry Holt & Co., 1881.

North, J. D. *The Measure of the Universe: A History of Modern Cosmology.* New York: Dover Books, 1965.

———. *Stonehenge: Neolithic Man and the Cosmos.* New York: HarperCollins, 1996.

———. *The Fontana History of Astronomy and Cosmology.* London: Fontana, 1994.

O'Malley, Michael. *Keeping Watch: A History of American Time.* Washington, D.C.: Smithsonian Institution Press, 1990.

Quiñones, Ricardo J. *The Renaissance Discovery of Time.* Cambridge, Mass.: Harvard University Press, 1972.

Reeves, Thomas C. *Gentleman Boss: The Life of Chester Alan Arthur.* New York: Alfred A. Knopf, 1975.

Robbins, Keith. *Nineteenth-Century Britain: Integration and Diversity.* Oxford: Clarendon Press, 1988.

Russenholt, E. S. *The Heart of the Continent.* Winnipeg: MacFarlane Communications, 1968.

Savitt, Stephen, ed. *Time's Arrows Today.* New York: Cambridge University Press, 1995.

Schama, Simon. *Landscape and Memory.* New York: Alfred A. Knopf, 1995.

Schivelbusch, Wolfgang. *The Railway Journey: The Industrialization of Time and Space in the 19th Century.* Berkeley: University of California Press, 1986.

Server, Dean. *The Golden Age of Steam.* New York: Todtvi, 1996.

Seward, William H. *The Works of William H. Seward.* Edited by George E. Baker. 3 vols. originally published 1853; reprinted New York: AMS, 1972.

Skelton, Oscar D. *The Railway Builders.* Brook & Co., 1916.

Smoot, George, and Keay Davidson. *Wrinkles in Time.* New York: Avon Books, 1993.

Smyth, C. Piazzi. *Teneriffe, an Astronomer's Experiment, or, Specialities of a Residence Above the Clouds.* London: L. Reeve, 1858.

Sobel, Dava. *Longitude.* New York: Penguin Books, 1996.

Stephanson, Anders. *Manifest Destiny: American Expansionism and the Empire of Right.* New York: Hill and Wang, 1995.

The Study of Time. 4 vols. Various editors. Berlin: Springer-Verlag.

Theroux, Paul. *My Other Life.* Boston: Houghton-Mifflin, 1996.

Thomson, Don W. *Men and Meridians.* 3 vols. Montreal and Kingston: Queen's Publisher, 1967.

Thomson, Malcolm M. *The Beginning of the Long Dash: A History of Timekeeping in Canada.* Toronto: University of Toronto Press, 1978.

Thoreau, Henry David. *Walden and Other Writings.* New York: Bantam Classics, 1981.

Time and Its Mysteries. 4 lectures. New York: New York University Press, 1940.

United States Department of Transportation. *Standard Time in the United States.* Washington: July 1970.

Van Deusen, Glyndon G. *William Henry Seward.* New York: Oxford University Press, 1967.

Vargas Llosa, Mario. *A Writer's Reality.* Boston: Houghton-Mifflin, 1991.

Vatsyayan, S. H. *A Sense of Time: An Exploration of Time in Theory, Experience and Art.* Delhi: Oxford University Press, 1981.

Ward, Mrs. Humphrey. *Lady Merton, Colonist.* 1910.

Warner, Brian. *Charles Piazzi Smyth: Astronomer-Artist, His Cape Years 1835-45.* Cape Town: University of Cape Town Press, 1983.

Zwart, P. J. *About Time.* New York: Elsevier, 1976.

Index